Combat Robot Weapons

Chris Hannold

McGraw-Hill

New York   Chicago   San Francisco   Lisbon
London   Madrid   Mexico City   Milan
New Delhi   San Juan   Seoul
Singapore   Sydney   Toronto

The McGraw·Hill Companies

Cataloging-in-Publication Data is on file with the Library of Congress

Copyright © 2003 by Chris Hannold. All rights reserved. Printed in the United States of America. Except as permitted under the United States Copyright Act of 1976, no part of this publication may be reproduced or distributed in any form or by any means, or stored in a data base or retrieval system, without the prior written permission of the publisher.

1 2 3 4 5 6 7 8 9 0 DOC/DOC 0 9 8 7 6 5 4 3

P/N 142201-3
PART OF
ISBN 0-07-142200-5

The sponsoring editor for this book was Judy Bass and the production supervisor was Pamela A. Pelton. It was set in Melior by Patricia Wallenburg.

Printed and bound by RR Donnelly.

McGraw-Hill books are available at special quantity discounts to use as premiums and sales promotions, or for use in corporate training programs. For more information, please write to the Director of Special Sales, McGraw-Hill Professional, Two Penn Plaza, New York, NY 10121-2298. Or contact your local bookstore.

 This book is printed on recycled, acid-free paper containing a minimum of 50 percent recycled, de-inked fiber.

To my Mom and Dad for pressing me.

Thanks to the following...

Jon Autry

Jason Bardis

Jeff Cesnik

Dan Danknick

Lawrence Feir

Tony Hall

Eric Koss

Mike Morrow

Brian Nave

John Reid

Terry Talton

Will Tatman

Contents

Introduction

Since becoming involved in the sport, I've seen what seems to be every type of weapon used in fighting robotics. There are spikes, spears, wedges, hammers, lifters, flippers, spinners, projectiles, smothering weapons, entangling weapons, flame throwers and liquids. Some of those are allowed in the combat rules. Some are not. Other weapon types may still be unused to this date. Those are the ones I look forward to.

In *Combat Robots Complete* I could not dedicate the amount of space to robot weapons, that I wanted to. This book will concentrate weapon systems. In it we will discuss how they are constructed, how they work and how effective they are when used against each other. Along the way, several veteran robot builders will give their opinions on their favorite type of weapon.

As in my first book, I claim that fighting robots are all about balancing your ideas with your capabilities. Robot weapons are no different. Building a robot that can actually crush an opponent could very well prove impossible. Your opponent is going to be strongly built and you have a weight limit that must be taken into account. You have to ask yourself how you can fit that huge hydraulic crushing mechanism into the weight class. How will you fit all the steel supporting struc-

ture into the weight class? All this is not meant to discourage you. It is meant to prepare you for certain realities that come along with the experience of building the ultimate robotic destruction machine. Once you realize that certain destruction of your opponent usually comes only by way of research, hard work and lots of failures, you will be on your way toward building a machine that can cause that destruction.

There are really two headings that weapons can fall under: legal and illegal. Different competitions allow some differences in weapons. One competition may allow flame throwers and liquids while another competition may ban them. None allows firearms or explosives. Because of arena strength, some competitions will limit the dangerous spinning weapons to the lower weight classes or ban them all together. The point is that you have to know the rules of the competition you will enter in order to build a machine that will be allowed to compete.

Every major competition publishes its rules on the Internet. Unfortunately, some competitions seem to alter their rules between every event. This is usually done because of some safety concern that originated in the previous competition but it does keep the builders on their toes, keeping track of specific design challenges. Fortunately, a group of event promoters is banding together to write a simplified rule set document. Once this group is finished, the plan is to have all the organizers run their events according to that document, with minimal changes, so that it is easier to build and compete in several different competitions. Until this project is finished, I leave learning the rules of the different competitions to you. This book will examine the myriad weapon systems allowed by Battle Bots, Robot Wars, and competitions like BotBash and NC Robot StreetFight, which allow similar weapons.

You can download each set of rules on each competition Web site. Here are four Internet addresses for the competitions mentioned above:

- North Carolina Robot StreetFight—http://www.ncrsf.com

- BotBash—http://www.botbash.com

- Robot Wars—http://www.robotwars.com

- Battle Bots—http://www.battlebots.com

Don't forget to check out the Robot Fighting League at http://www.botleague.com.

1
History First

It seems every author who writes a book about combat robots thinks it necessary to include an explanation of how the sport came to be. I did not do so in my first book, which was about how to build the bots. This book is about the weapons of those robots. Nevertheless, you might want background on me, the robots I've built, the people I've met, and the competitions that I've been to. Rather than rehash a timeline of events, I'll drop in a bit of what was happening in the world of combat robotic sports while I was working my way up to becoming a veteran builder and competition organizer.

The Beginning

I've been building some form of robot for more than twenty years now, which puts me back to right before the Internet came into being. Back then, I would take appliances apart to get the materials I needed to build robots. Most of the stuff I built was supposed to be in the autonomous class of robots. An autonomous class robot runs by itself, making decisions on where to go and when. I was one of two people in my entire school who was building robots. There seemed to be absolutely no local outside enthusiasm about my hobby. The only hint

of a world of robotics was through magazines like *Radio Electronics*. Even then, there weren't many articles on how to build them.

In eighth grade I was lucky enough to have a science teacher who did everything he could to get his students involved in the class. If the topic was astronomy, he took the class to a planetarium and to watch the stars and planets through his own telescope. One of the coolest things I've ever seen is a tiny white spot with rings around it. It was Saturn, through a telescope. If the topic was solar power, he built an eight foot-wide parabolic dish that focused the sun's rays onto a collector and heated water. The coolest part of that was the automatic sun tracking capability. The dish would search for and find the sun. then it would track across the sky, following the sun and making sure it had the best possible angle to collect the most rays.

Before I met this teacher, everything I knew about electronics was confined to discrete components and circuits. To me, the computer was a box that had some games installed. I never knew what it took to make a computer control the world or even that it could. That changed when the teacher, Joe Teeter, built a machine that could grade tests. Just like the machines that grade the SATs or whatever other standardized tests you might take, this machine read which circles had been colored in with a pencil. A TRS-80 computer built by Radio Shack controlled it. Mechanical, electronic, and controlled by a computer, this machine was the closest thing I'd ever seen to a real live robot. The only thing missing was movement. With the help of my teacher, the support of my Mom and Dad, *Radio Electronics* magazine, *Popular Electronics* magazine and the first edition of Forrest Mims' *Engineer's Notebook II*, I was on my way to building real robots.

As I went through high school, my attentions turned a bit toward cars and girls. My grandfather owned a service station where he and my uncle were the mechanics. I learned a lot from them when it came to cars. In high school, along with advanced math and physics, I took auto mechanics. At one

time I almost decided to become a mechanic. However, it seemed that robotics and computers were stuck in my mind.

Shortly after college I saw a Discovery Channel show about fighting robots. At the time I was searching for a career close to home and barely had money to pay the bills, much less build a fighting robot. I continued to build small robots from spare parts and inexpensive components. There seemed to be nothing greater than the feeling of watching one of my creations come to life and do what I intended.

A couple of years went by and I did find the job I wanted. The Internet was just getting popular and the newspaper I worked at wanted me to set up their Web site. That was back in the days where you could type in www.(whateverterm).com and still possibly get a "404 File Not Found" error on your browser. While doing some research on the Net, I came across a Web site about fighting robots. It of course contained pictures and links to a few other sites with more information. That's when I got hooked. One of the most helpful sites I came across back then was that of Team Delta (www.teamdelta.com). The writer of the site, Dan Danknick, is one of the most creative builders I've ever met. The site still exists. In it, you'll find his creations, lots of information about the robot combat world and lots of excellent products built solely for combat robots. Dan furthered my interest in autonomous robots as well as combat robots. He's the one who got me started using the PIC microcontrollers by Microchip (www.microchip.com). These little controllers are computers on a single chip. They are used in several circuits that decode the remote control signals for combat robots and can be easily used for autonomous robot control.

At the time, it didn't matter that the only robot-fighting event in existence was held in California and I lived in North Carolina. I had to build a machine that was remote controlled and could tear things up. It was about the same time that Marc Thorpe, inventor of robot combat, was getting some backing to help fund his California-based events. The first few events went off without a hitch and without me. My first robot VOID,

Figure 1.1

VOID, my first combat robot.

shown in **Figure 1.1**, was not finished. In fact, I never really finished VOID at all. I got it to the point of running around. The next year, Robot Wars was broadcast over the Net and I was able to watch it as it happened. I still hadn't finished VOID. For some reason, robotic combat had not completely consumed my life at the moment, and I had other priorities like bills and going out on Saturday night.

Starting Over

Watching the fights on the Net was an eye-opener. I realized that my design for VOID would not do well. It was too high off the floor and the weapon was too complex to operate effectively. Not to mention that it wouldn't be very effective. So, after spending a couple of thousand dollars on sprockets, bearings, batteries, speed controllers and a radio, I changed designs and started over. This time I was building for a small competition that was part of a larger science fiction convention called Dragon Con in Atlanta. Fifty pounds was the highest weight limit. I sat down with pencil and paper trying to draw a small, fifty-pound box that would use the same batteries and speed controllers that I had tried to use in VOID. I finalized my design and called it XAK (**Figure 1.2**).

Figure 1.2

XAK, my second combat robot.

The Robot Battles of Dragon Con were the first competition I ever entered. My brother, who had never been interested in robotics, came along to take some pictures. XAK was defeated by a large wooden bot when I had problems with the radio signal. XAK ran off the platform and almost struck the man running the show. Around the same time, Thorpe was having problems with his backer and the Robot Wars event was canceled because of lawsuits. Robot Wars ended up in the hands of the backer and was taken to England, where it has flourished ever since. Shortly after the lawsuit ended, BattleBots was created by Trey Rosky and Greg Munson. BattleBots pretty much resurrected the sport of robot combat in the United States.

Going to BattleBots

BattleBots held a couple of events, including one that was broadcast on pay-per-view TV. I could not attend either of these because I still had not been consumed with robot fighting and insisted on paying the light bill and was still somewhat wrapped up in autonomous robotics for things like the Trinity College Fire Fighting Home Robot Contest. However, the moment of truth came when BattleBots announced their

Figure 1.3

RipOff, my
entry into
BattleBots™.

next competition was to be held in Las Vegas. I had been designing a vertical spinner robot for a while now and decided this was the time to go to the big show.

RipOff (**Figure 1.3**) was my concept of a robot with a vertical spinner weapon. I believe it was one of the first designs to mimic the ever-popular robot from Jim Smentowski called Nightmare. RipOff was similar because of the weapon, a vertical-spinning disk. In this case the spinning disk was a 28-inch saw blade from an old wood mill. I did not count on the cutting ability of the blade. Instead, I hoped the big teeth would grab some piece of the opponent and rip it off. I gave the robot its name for that reason and because it was a clear ripoff of the Nightmare robot.

My brother was getting hooked by now and came in to help build the bot when it got close to the deadline. Unfortunately we were ill prepared. In the process of building, testing, and competing we blew up five speed controllers and the robot never moved off the red square of the battlebox until our opponent, Robert Everhart's Atomic Wedgie, slammed into us and knocked us around the ring. Later I figured out what I was doing wrong. Vegas was a blast. It was just too bad that BattleBots would never hold another event there. They decided to hold the rest of their events in San Francisco. Now that Comedy Central has decided not to renew the BattleBots TV contract, maybe Trey and Greg will bring the event back to Vegas.

Going to Robotica

Soon after returning from BattleBots, The Learning Channel (TLC) announced their new robot combat show called Robotica. This show was to be something completely different from either BattleBots or Robot Wars. It would have three different types of obstacle courses and a fight at the end. The twenty-four competitors would be separated into six groups of four. Each group of four got its own episode. Within the groups, two robots would face each other in the obstacle courses for the chance to face the winner of the other two bots in the fight to the finish. The winner of the episode would continue against the winners of the other episodes. They continued this until three bots were in the final fight to the finish. The overall winner received $10,000.

Somehow my application was accepted to be one of the twenty-four bots to compete on Series 1. The producers were looking for a mix of experience, and my brother and I had been gathering parts for a 100-pound bot but we had not started to build it yet. When we were told that we had been accepted, we reviewed our design and decided that it wouldn't work for the show. The weight limit TLC had set was 220 pounds. Also, there were speed bumps that we would not be able to get over and heavy objects that we would not easily be able to push out of the way.

We quickly drew up plans for HandsOff! (**Figure 1.4**) on the floor of the shop. The producers wanted a more technically savvy show than what BattleBots was doing so they wanted video tapes of nearly everything. They even flew a guy out to our shop to do an interview and to tape some preliminary driving and weapon tests. Since they wanted something technical, I decided to build my own gearboxes. The main weapon was again a spinner. This time it was a horizontal spinning bar. The producers seemed to be happy with what they had seen. Three free tickets to Hollywood, paid hotel, and shipping expenses later, we were at the competition.

I saw several familiar faces from the BattleBots competition and even more new faces. We made friends with nearly every-

Figure 1.4

HandsOff! during construction.

one. Mike Morrow of Team Juggerbot (www.juggerbot.com) had been following our build reports on my Web site. He even brought us a "Hands Off" tag. Mike's robot, JuggerBot, was a four-wheel drive bot that was low to the ground and had lots of pushing power. It is shown in **Figure 1.5**. It had no active

Figure 1.5

Mike Morrow's JuggerBot. (Photo courtesy of Mike Morrow and Team JuggerBot)

weapon but JuggerBot's technology could have taken first prize if that was one of the criteria. The wheel pods were replaceable in case of any damage. They even had a method of downloading the temperature of the motors and batteries; the current that had been used was readily available too. Mike and his team ended up in the fight to the finish going after the Robotica winner title. They lost due to what I think was a small driving error. Many wins and losses come down to just that.

This was the same time we met up with Jeff Cesnik and the crew of Team Suspect (builder of Kritical Mass). Getting to know these guys was a blast and made the long hours of waiting seem bearable. Kritical Mass, shown in **Figure 1.6**, was basically a titanium box with a wicked vertical spinning weapon. A six-wheel-drive system put the power to the floor. The titanium was mounted so that it touched the floor all the way around, making it next to impossible to get under the bot. The spinning blade was milled from a solid chunk of tool steel. When KM was on the fighting platform with the weapon running, it tore up part of the set. This led the producers to modify the fencing to make it stronger. Obviously they hadn't planned on robots that were so powerful.

I mentioned that there were long hours of waiting. There were four different sets. The racetrack was the first set and was pretty much painted on the floor. Moving the barriers around didn't take too long. However every other set was completely different

Figure 1.6

Jeff Cesnik's Kritical Mass. (Photo courtesy of Jeff Cesnick and Team Suspect)

and had to be pushed around by hand into the same spot. The show had never been done before, so the crew did not have much idea of what it would take to move each set. This meant that the builders and audience were made to sit around and wait. The schedule was grueling. We were there at about 7:30 AM and many times didn't get to leave until 2:00 or 3:00 AM. There were many more problems that needed fixing, but all that was forgotten when we got to set up in the pit area.

We were pitted against Andrew Lindsey's Ram Force, a strong, boxlike robot. We did well enough in the Racetrack but they beat us on points. There were a few collisions, one that broke out a piece of their polycarbonate armor. We didn't know it but there was a problem with one of our gearboxes. We had used roll pins to secure the gears to the shafts. During one of the collisions in the racetrack, two of the pins came loose. One pin fell out in the pit area without our knowing it and caused us to have a very poor performance in the Labyrinth. We repaired that pin, but in the rush did not notice the other pin was about to fall out as well. The second pin dropped onto the cart used to carry the bot to the Gauntlet. The lost pin cause a serious lack of power on the right side of the bot and it ran right into the set wall of the Gauntlet. The spike on the front of HandsOff! penetrated the plywood set. HandsOff! is shown completed in **Figure 1.7**. Because the spike was tapered, as it went into the plywood, the rear of the bot was raised off the ground enough so that the wheels would only spin. There was nothing we could do as we watched Ram Force win with an overall score of something like 180 to 10.

Going to Robot Wars

Never taught to be quitters, my brother and I kept at it. Right before leaving for Hollywood, Robot Wars (in England) announced that they were looking for American competitors, and I sent in an application. Very shortly after returning from Robotica, we heard that we'd been accepted to compete in England in only a few months. This time we looked at

Figure 1.7

HandsOff! finished.

HandsOff! and decided there were a few design flaws that needed to be changed. For one, the spinning weapon would hit nothing that wasn't 15 inches tall. There's a reason for that which involves a misread of the rules for Robotica. We needed to change that weapon so that it would hit an opponent that was 4 inches or taller. We also wanted to get rid of the roll pins in the gearboxes.

Changing the spinning weapon meant a radical change to the frame of the bot. It no longer looked the same, so a new name was in order. The Six Million Dollar Mouse (**Figure 1.8**) was born. We changed the weapon, beefed up the motor that was spinning it, changed from roll pins to key stock, changed the paint scheme, and added a self-righting mechanism that looked like a mouse tail. We also changed the gear reduction ratio to get more speed. It had plenty of torque already. So we crated the Mouse and headed to London, England for the taping of the first Robot Wars: Extreme Warriors. It was to be shown on TNN later that year.

The flight to London was a long eight hours. Once we got there, we grabbed a taxi and got to the hotel. I made some new

Figure 1.8

Six Million Dollar Mouse.

friends, including Jerome Miles (builder of BattleBots' Sublime and Robot Wars' Unibite) and Brian Nave (builder of BattleBots' Phrizbee, Phrizbee Ultimate, and Robot Wars' Revolutionist). Brian and his team, Team Logicom (www.teamlogicom.com), specialize in robots with spinning weapons. Phrizbee and the Revolutionist, shown in **Figure 1.9**, are both "hockey puck"-style robots. The entire shell rotates around a stationary body. The shell is made of steel and has lots of sharpened steel teeth welded onto the outside.

Figure 1.9

Brian Nave's Revolutionist. (Photo courtesy of Brian Nave and Team Logicom)

The shell spins very fast and the robot can be steered to attack the opponent. When it hits, the teeth take lots of little bites out of the opponent. This type of robot creates a lot of destruction and is very tough to beat if well built.

Team Suspect brought Manta (**Figure 1.10**) a redesign of Kritical Mass with an even larger spinner blade. The Six Million Dollar Mouse lost its first fight against four other bots in a judges' decision. This was okay since they had plenty of fighting planned for everyone. We entered the North vs. South contest. We defeated two opponents before being whipped by Manta in the final round. Next we entered the US vs. UK fight and were beaten by Pussycat. Talk about ironic.

Starting My Own Competition

There were no problems with the Mouse in London. When we packed it up, it was running just as well as when we sent it over there. This was good because as soon as we got back to the states, we were planning the first NC Robot StreetFight for July

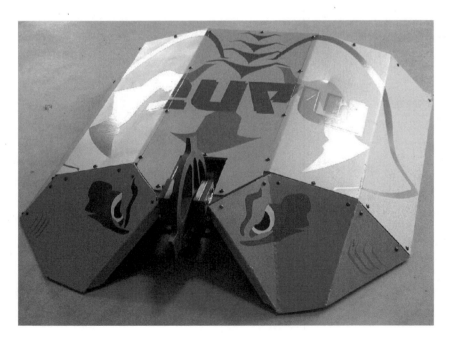

Figure 1.10

Jeff Cesnik's Manta. (Photo courtesy of Jeff Cesnik and Team Suspect)

2001. When I first thought of it, I had only planned on getting a few builders together to have some fun out in a parking lot. By the end of May, we had about fourteen bots from seven states signed up. I couldn't stand the thought of bringing people from out of state to a fairly unorganized get-together, so we upgraded our plans and rented the county fairgrounds.

A sponsor paid for lunch for the builders and the crew (**Figure 1.11**). The Mouse was the only 220-pound bot in attendance so I bumped it up into the Super Heavyweight class where there were two other opponents. Matt Ulrey of Team CUAD had brought CUAD the Crusher (seen on BattleBots). Eric Koss of Team Triborg (www.kosscomputers.com) had brought Isosceles (**Figure 1.12**), which he had taken to the previous BattleBots competition. The Mouse was out-weighed but survived the attacks. With the help of several relatives and a few friends, we held a safe and fun robot combat event. At the time, it seemed like a lot of work but everyone had a good time. So we decided then and there to have another one.

Back to Hollywood

Before planning for the second NCRSF got well underway, the producers of TLC's Robotica called me. They were planning

Figure 1.11

The crew of NCRSF #1, July 28, 2001.

Figure 1.12
Eric Koss'
Isosceles.
(Photo courtesy
of Eric Koss and
Team Triborg)

on taping two more events in the Fall and needed us to build a machine as an alternate. They only gave us three week's notice but MiniRip (**Figure 1.13**) was born within a week of the phone conversation. MiniRip was a conglomeration of parts from VOID, RipOff, HandsOff! and the Six Million Dollar Mouse. It had a bulldozer blade and a small spinning hammer disk. This was our first four-wheel-drive robot too. We finished MiniRip in record time and crated him up to be

Figure 1.13
MiniRip, winner
of episode 1
of Robotica
series 3.

sent to Hollywood. We had a feeling that we were no longer just going to be alternates. Some robots fell off the list because of the problems of September 11, and sure enough, they needed us for the show.

We were pitted against Hammerhead. This bot's builder had been to BattleBots a few times and the robot was solid. It had a lifting wedge and was articulated in the center so that it could go over bumps and debris without many problems. The show format had changed a bit and the crew knew what it was doing this time. Instead of being on the set for nineteen hours, like during Series 1, we were on the set at between 8:00 and 9:00 AM every morning and off the set around 6:00 every night except the last night. This gave the builders a chance to hang out together and take a look at Hollywood nightlife.

We tied Hammerhead in points on the redesigned Gauntlet and beat him on points in the redesigned Labyrinth. This meant that we would face one of the other two opponents in our episode. A robot named Pfshht! by Tonya Bingham was to be our fight-to-the-finish opponent. It was by coincidence that I had helped Tonya and her husband design Zero when they found out they had been accepted to compete in Robotica Series 2. Zero did well in that series, and they were invited to modify the robot and compete as a new team in Series 3.

In the fight to the finish, the robots are placed on a platform about 8 feet high. Fencing a little more than a foot tall surrounds the platform. The fence drops out of the way after 60 seconds of fighting. Once the fence drops, the object is to push your opponent over the edge into a field of steel spikes.

Our fight to the finish with Pfshht! was a short one, at about 65 seconds. One head-on hit with MiniRip's bulldozer blade and their polycarbonate wedge was ripped from the body and tossed aside. Pfshht! was also turned upside down in that collision. This made it difficult to drive, and just as I was lining up to push Pfshht! over the edge, it went by itself. We had won our episode, a bit of cash, a shiny brass Robotica medal, and a spot in the finals.

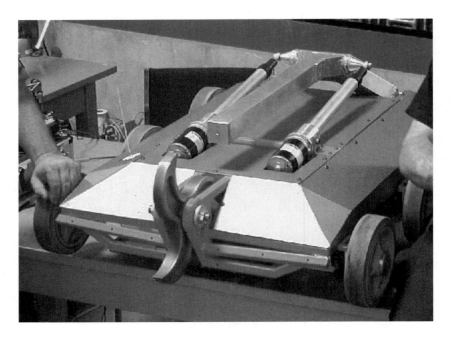

Figure 1.14

Jeff Cesnik's Ultra Violence. (Photo courtesy of Jeff Cesnik and Team Suspect)

In the finals we went up against Jeff Cesnik and his new robot, Ultra Violence (**Figure 1.14**). It was another redesign of his original Kritical Mass bot with an even fiercer spinning blade. Jeff had used that nasty blade to turn a beautiful wooden bot into toothpicks just the day before. With six drive wheels, Ultra Violence was a bit more controllable than MiniRip. However, at one point, we had both become stuck on some debris but UV was able to clear the forest of glass and gain the bonus points to beat us. We weren't out completely. The producers introduced a new aspect to the show. There was a wild card slot in the final four that would be filled up by the bot with the best time in the Gauntlet. Unfortunately the Juggerbot team and the Jawbreaker team tied in points in the Gauntlet, thereby canceling our hopes of snagging that wild card slot.

Down to Florida

Being back home and already wanting some more robot action, I started designing the arena for the next NC Robot StreetFight. We ordered the steel, a new welder, and a new

band saw and began construction of the 32x32-foot-square arena. In the mean time, another East Coast event was born, the SECR Farm Fight (www.secr.org). This event was held on, you guessed it, a farm. My brother and I took a weekend to throw a 30-pound spinner robot together and called it Mighty Mouse (**Figure 1.15**). We trucked down to Florida without weighing or testing the new Mouse. That was a problem. When we got there, the bot was overweight without armor or batteries. So we slapped on some spare polycarbonate and stainless steel. After that, we had problems with the radio and were defeated in the 60-pound weight class.

Figure 1.15

Mighty Mouse.

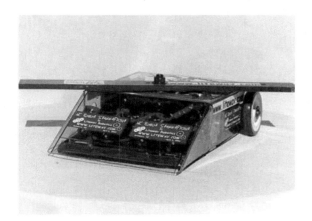

NC Robot StreetFight 2

Returning home yet again we dove into arena construction for our next event. The Farm Fight had about twice as many entrants as our first NCRSF so we knew interest was growing for competition on the East Coast. We skipped the next call for roboteers for Robot Wars Extreme Warriors in order to complete our arena. We finished it the week before the event was to happen. On April 6, 2002 the second NC Robot StreetFight had fifty-one robots signed up from fourteen different states. What a sight!

The second NCRSF was the first one to be opened to spectators. Once again, relatives and friends came together to pull

Figure 1.16

The crew of NCRSF #2, April 6, 2002.

the event off without a hitch (**Figure 1.16**). The main difference was that we gained recognition in the robot combat community and this, in turn, gained us sponsors. More sponsors meant more prizes for the competitors. The sponsor, competitor, and spectator turnout convinced me to organize the third NCRSF.

Combat Robots Complete

Going back a little, after returning from Robot Wars Extreme Warriors, I decided to write the book *Combat Robots Complete* (**Figure 1.17**). It took a year and a half to finish all the research and write. During that time, I built another 30-pound bot specifically for that book and called it Dagoth (**Figure 1.18**). The name comes from a *Conan* novel. Dagoth made his debut as a 30-pound wedge and spike bot at the third NC Robot StreetFight (www.ncrsf.com).

NC Robot StreetFight 3

The third NCRSF was held on December 7, 2002. My brother and I constructed a new roof for the arena of steel angle and two

Figure 1.17

Cover of *Combat Robots Complete*.

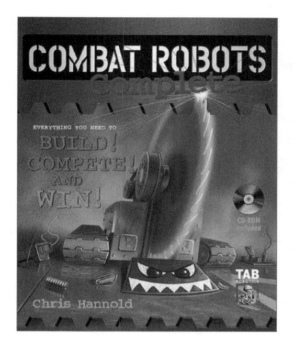

Figure 1.18

Dagoth.

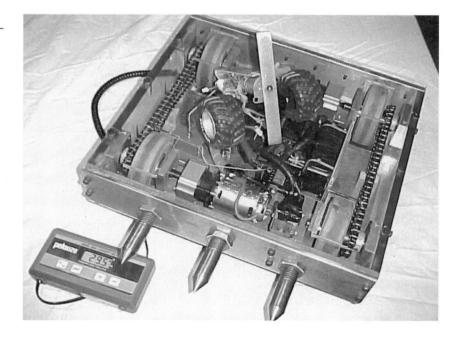

layers of plywood with a total thickness of 1-1/2 inches. The major news about this event is that North Carolina had its worst ice storm ever, just three days before. This ice storm left one and a half million people across the state without power. This included me, all but one of my crew, and the venue itself. I hauled my computer, which held the mailing list for the registered competitors, across town to the only house with power and sent out an announcement that we were in the dark and the best guess about the power was that it wouldn't be back on for a week. I did not call off the event but I tried to let everyone know that we'd most likely only be able to fight during daylight hours. The irony was that this was still a one-day event, I had sixty-one bots registered, and we were approaching the shortest day of the year. To my amazement and as a testament to the robot builder community's devotion to the sport and its organizers, many builders volunteered to bring generators so that we could charge batteries and have arena lights. Without knowing that power was restored late Friday afternoon, we had almost fifty bots show up. Some of them came in from as far away as Ohio, Vermont, and Florida. Dan Danknick of Team Delta (www.teamdelta.com) was an event sponsor this time around. He even flew in from California to help out. As I said, the power came back on Friday. We got started on time Saturday morning even though it was about 28 degrees and there was no heat in the venue. There were seventy-two paying spectators all day long, thanks to the ice and the power problems. Still, everyone seemed to have a good time. I have to say, I've never met a finer group of people than those in the robot combat community. **Figure 1.19** shows a picture of the improved arena. I'm sorry to say I don't have a picture of my crew.

Building the Village Idiot

I guess the thing I am most proud of with my initial build of Village Idiot, is that I made it in my garage with little more than a hand drill and a Dremmel, but it was still competitive. I learned most of what I knew before starting on the robot by reading on the Internet. Sites like Jim Smentowski's robotcombat.com, Chris Hannold's

Figure 1.19

Arena picture.

litewav.com, and Derek Young's automatum.com made me think, "Hey, I can do this too!" and provided me with the ground info to get started.

The first Village Idiot design process consisted of trying to come up with something different, but hopefully effective. I feel like I was lucky enough to find a good mixture between the two. Construction was as informal as you can get. I basically built and engineered parts as I needed them and solved problems like "how am I going to get this to fit there" on the fly. This method of building made for a LOT of unplanned frame modifications. As I have gained experience in competitions, and in building, I have dramatically refined the steps I go through in creating a new robot.

One of the most important lessons I have learned thus far in bot building is that thorough planning will save you both time and money. In the most recent build of Village Idiot, I completely designed the frame and base plate with Computer Aided Design (CAD) software. By rendering a complete 3D model of my robot, I was able to see BEFORE welding a piece into place, if everything would fit and how much it would weigh.

The most dramatic change in Village Idiot's design in the current rebuild is the move from an electric weapon motor to that of an internal combustion engine (ICE). I made the decision to change weapon motors for several reasons. In no particular order they are:

- Anyone who has attended a live event and experienced the thrill of watching a robot compete with a screaming two-stroke engine will tell you, "The crowd LOVES a gas engine."

- Gas engines have a very impressive power to weight ratio, and they do NOT require batteries to provide power. A few ounces of fuel vs. several pounds of batteries is a very attractive perk when you want to add mass to weapons, and armor to your robot.

- The weapon disc assembly on the new Village Idiot weighs almost 30 lbs, compared to the previous version that weighed ~14 lbs. This weight bonus was partially achieved in being able to remove batteries and use a lighter motor without losing much power.

- I wanted to learn something new. The sport of robotic combat celebrates learning, and the people you will find that are involved with the sport are addicted to acquiring new skills and honing the ones they already posses.

Unfortunately, there's an equal amount of "down-sides" to using a gas engine:

- They have proven to be less reliable in combat vs. their electric brethren.

- The potential to "stall" the engine is always present. Without the added weight and complexity of an electric start, your weapon is out of commission in the event that your engine stalls.

- On the whole, it is far simpler to engineer the robot to securely contain the electronics and parts for an

electric motor vs. the difficulty of securely containing an internal combustion engine, its fuel, the servo for the throttle, etc...

Like many of the choices people are faced with in robotic combat, there are pros and cons to either type of power plant you choose for weapon locomotion. Builders usually end up making the decision that they feel will best suit their designs needs, and that they will get the most enjoyment out of building.

Jon Autry, Team RV (www.robot-village.com)

Figure 1.20

Village Idiot. (Courtesy of Jon Autry and Team RV)

Back to Florida

The last competition I attended before getting this book done was called Battle Beach (www.battlebeach.com) and took place in March 2003. Brian Nave, of Team Logicom, is the organizer of this very cool robot fighting event. It took place in the last week of Bike Week in Ormond Beach, just a few miles away from Daytona. None of my crew could make the four-day

event, so I went by myself. Unfortunately for me, Brian's bot, Phrizbee, had ripped the wheels, spinner weapon, and bulldozer blade off of my heavyweight bot MiniRip during the last NCRSF. I had to get it going again.

We wanted MiniRip to be Phrizbee-proof this time around. I'm not saying that it was, just that it might have been. We won't know until we meet up in the ring again. Brian didn't fight his own robots. He had plenty of other things to do as the organizer of the event. I can certainly understand. Anyway, Greg and I wanted to add a wheel guard. We also needed to fix some holes in the bulldozer blade. It also turned out that one of the Phrizbee hits had bent a solid steel, 1-inch diameter axle. When that happened, it also tore up a steel bearing housing, severely bent the aluminum side wall of the bot, and broke two teeth off a sprocket.

Rather than spending a lot of money on new parts, we decided to hammer out our problems, literally. With a bit of patience, a big hammer, and an anvil, we reformed the bearing housings, straightened some of the body panel and part of the bulldozer blade. We couldn't use brute force on the bent axle so we made up a new one on the lathe. Part of the body was bent so badly that we had to use a plasma cutter to remove it. Then we cut a new piece of material and welded it back into place. I fixed the sprocket teeth with a welder and a file to reform them. After the major repairs were done, we stripped all the paint off and gave it a new coat.

The bulldozer blade was only held on by six, 1/4-inch bolts in the original design. This was strong enough to withstand the bricks, paint cans, and glass of the Robotica television show but it couldn't stand up to a Phrizbee attack. We welded on two pieces of steel angle and drilled holes so that it would bolt onto the bot down its entire length. The new wheel guard was to be partially held in place by bolting it to the bulldozer blade and onto the rear end of the bot. The guards themselves were made from steel "C" channel that was 5 inches wide and 5/16-inch thick. **Figure 1.21** shows the new look of MiniRip.

Figure 1.21

MiniRip after repairs.

You might ask how I added all this steel to the bot since there is a weight limit. Well, the answer is that the spinner weapon was too severely damaged to be used again. The bearings and axle were fine, but the 3/8-inch thick disk was bent and the steel housing was basically crushed. The entire spinner mechanism weighed about 35 pounds, and we removed all of it. We also moved the bulldozer blade to the other end of the bot. Doing this changed the center of gravity around. Sometimes this is a bad thing. However, in this case, it was a great thing. The bot now handled excellently and kept the blade down on the ground where it needed to be. In past events, the bot was popping wheelies and was a general pain to drive.

When I got to the event I found a pit table next to Jon Autry and Adam Baxter. The guys behind us were Jim Klein and Doug Groves. You'll recognize all these teams as winners at the December 2002 NC Robot StreetFight. Once I got unpacked and settled in, I realized that the bot builder community hadn't changed a bit. It was still one big happy family that liked hanging out and having fun together. This fact is still my favorite thing about combat robots.

Brian's event concentrated on the "show" aspect of fighting robots more than any privately run event that I've attended.

He had a full production crew running lights and cameras. He had Roy Hellen and Brett "Buzz" Dawson doing color commentary on all the fights along with driver interviews. Roy was the guy wearing shiny armor and handling bots at Battlebot's events. Buzz is the captain of Team DaVinci. They built another BattleBot fighter called Moebius. Buzz is also host, along with Brian Nave, of a new robotics show coming out on Tech TV cable channel. Roy and Buzz definitely entertained the crowd and did an overall incredible job as announcers.

MiniRip did very well at the event too. We took fourth place. All our wins were by knockout. Since it was a double elimination tournament, we had to have two losses to be booted. One loss was a judge's decision and one was a knockout. The judges did an excellent job since all the damage visible was to MiniRip. The aggression and strategy categories were close though. In our only knockout loss, we went up against a mean robot from Team Van Cleve called Tornado Mer. We put him in the loser bracket in one of our earlier fights and met up against him after our first loss. Tornado had a 55-pound spinning blade made of S7 hardened steel. In our first fight against him, we got him upside down after a few good hits and he was counted out. In our second fight against him, he took a huge chunk of steel out of our bulldozer blade. When that happened the forces jarred the batteries loose inside the bot and all the power cables pulled out of their connector. This killed our bot and we had to tap out.

Summary

Over the course of the past few years, my brother and I attended every competition that we could. BattleBots and Robot Wars have gained immense popularity all over the place. I've met some of the best people in the world and made some new friends. Building robots has given my family a chance to travel to places that we had never been and probably never would have. Small competitions are popping up all over the United

States too. It only makes sense that someone would want to get involved in the fastest growing sport in the world. It also makes sense that someone would want to see what kinds of weapons have been tried and what the best strategy is while using them. I've seen hundreds of robots and even more robot fights. The weapon systems these robots use vary with each incarnation of mechanical monster that takes the field of battle.

In the following sections of this book I will talk about what is legal and what is illegal in the major competitions and the smaller ones that are based upon their rules. We'll go over the different types of weapons in use today, how they are constructed, and how each one will fare against the others. We'll also talk about the mechanical and electrical aspects of these weapon systems.

2
Legal
versus Illegal

Each fighting robot weapon typically falls into one of several subcategories of weapons that are allowed in a robot fighting competition. As we discuss the subcategories, I will let you know if the weapon type is *not* allowed in a certain competition. The aim is to discuss the weapon types that are commonly allowed at all competitions similar to BattleBots, Robot Wars, and a fledgling league called the Robot Fighting League (RFL—www.botleague.com). The RFL is composed of many small events across the United States. One of its main objectives is to form a standardized rule set so that builders can follow it and compete in as many of the RFL events as possible with as few changes as possible.

Everyone dreams of mounting a giant flame thrower or an electromagnetic pulse generator on their fighting robot. The plain fact is that these are not allowed in most of the major competitions. The restrictions on weapons start with robot construction materials. The following information comes from version 2.1 of the BattleBots Technical Regulations, publicly available at the BattleBots Web site.

Restricted construction materials include:

- Any lead (Pb) metal must be used in a way so that your opponent can not damage it. (That rule alone will cancel any thought of using a conventional firearm.)

- Rigid plastic foams (PVC, polystyrene, polyurethane) should not be mounted so that your opponent can cause them damage.

- Permanent magnets (not the ones in the motors) can be mounted on the outside as long as they are glued or bolted in place.

- Expanding (pourable liquid) foam cannot be used to fill up all the empty cavities of your bot unless you can do it and still have access to all the parts for safety inspection and maintenance.

- Any magnesium sheet should be 1/8-inch or thicker. (Ever try to put out a magnesium fire?)

Forbidden construction materials include beryllium, mercury, boron, lithium, sodium, asbestos, depleted uranium, radioactive materials, unsintered metal powders, composite fibers like fiberglass and carbon that are not woven or bonded, metal shavings, wool, and decayable organic substances other than leather and wood.

You are permitted to use lasers, lights, and sounds on your robot but not without restrictions. Lasers are restricted to Class II and must be aimed at the floor so that they do not shine more than 4 feet from the robot. Sound is restricted in that you should not be able to measure more than 120 A-weighted decibels when 10 feet away from the bot. Black lights that emit 400nm wavelengths or shorter are forbidden. Any lights that are deemed "distracting" will not be allowed. That means you cannot have a search light mounted to your bot with your partner controlling it to shine in the eyes of your opponent.

The list of restrictions on construction materials seems large, but the fact is you can use almost anything else. The difficulty in building robotic weapon systems is not in finding materials. The difficulty is in finding and building a design that can withstand the extreme forces that these machines encounter during battle.

Spinners

The spinning weapon category, some of them shown in **Figure 2.1**, has its own set of restrictions to accompany the general construction restrictions. The first is really part of a general robot operation rule. That rule says that all weapons and drive mechanisms must come to a complete stop if the remote control transmitter is turned off. This must happen whether the robot is sitting still, running at full speed, or if the weapon is spinning at full speed. That ability is designed into the radio's fail-safe mechanism and should be set according to the radio manufacturer's documentation. Basically, the robot must power off by itself if contact is lost with the person running it.

Assuming you have the fail-safes set correctly and the robot and its weapons come to a stop if radio control is disrupted, your spinning weapon could possibly take a long time to come to that complete stop. The heavy mass spinning at high speeds has a high moment of inertia. That means it doesn't want to stop. That's also what makes spinning weapons so effective. I've seen large grinding wheels take twenty minutes to stop. I've also seen the rotors of large shears take two hours to stop

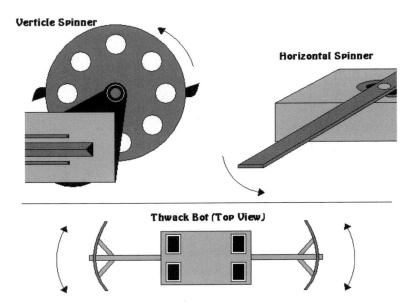

Verticle Spinner

Horizontal Spinner

Thwack Bot (Top View)

Figure 2.1

Drawing of various spinner-type weapons.

spinning. Most competitions require your weapon to stop spinning within a certain amount of time. Thirty seconds to one minute is the most accepted length of time to stop. Unfortunately you may be required to install an automatic braking system. This is not really difficult, but it does take up a bit of your weight allowance that could otherwise be used for armor, batteries, stronger motors, or whatever else you might think to add.

Pneumatic or Hydraulic Powered Weapons

You have your choice of what power source will make your weapon move. Lots of people use electric motors. Several people use pneumatic (air, CO_2, etc.) systems. Some people use hydraulic (oil) systems. Some use internal combustion engines (ICEs) to power their weapons. Electric systems are regulated throughout the rules. For ICE systems there are a few specific rules regarding the amounts and types of fuel that can be used, placement and armament of critical parts and, of course, the overall requirement that the systems shut down upon loss of radio signal.

Hydraulic weapons are typically powered by either an ICE system or an electric motor. A typical hydraulic weapon powered by an electric motor is shown in **Figure 2.2**. Therefore, that part of the hydraulic system must comply with the rules governing either the ICE systems or the electric systems. The regulations imposed on hydraulic weapons systems are fairly complex. They take up four of the fifty pages in the BattleBots Technical Regulations document. The rules range from an inane "hoses must be secured to prevent vibration" up to a noteworthy "hydraulic systems are not allowed in 60-pound robots."

Pneumatic systems are probably the most regulated of all weapon systems. A basic system is shown in **Figure 2.3**. Check with the competition rules for specific information. These are also some of the most versatile systems available to builders. Pneumatics can be used to swing a hammer, to lift or

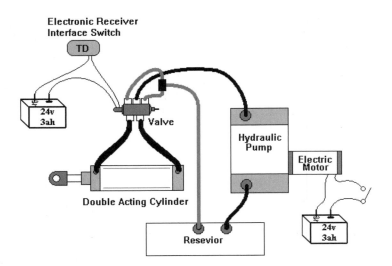

Figure 2.2

Hydraulic weapon system powered by an electric motor.

flip your opponent, to slam a spike into the opponent, to hurl a projectile at the opponent, and to flip the robot over if it gets upside down. I'm sure there are many more uses you can dream up. There are also multiple regulations ranging from limits on the gas capacity to isolating the system lines from any heat-generating bodies inside your bot (motors, batteries, etc.) The pneumatics section of the BattleBots Technical Regulations document takes up nine of the fifty pages.

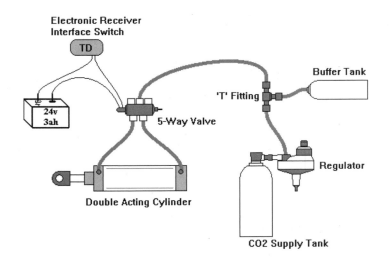

Figure 2.3

Basic pneumatic weapon system.

Overhead Hammers or Spikes

There are no specific restrictions on building an overhead attack weapon unless you plan to have the robot jump up and attack from the air. In that case, according to the BattleBots rules, the robot can jump no more than 6 feet in the air and for a distance of no longer than 10 feet across the arena in a single leap. The jumping bots are not allowed to damage the arena floor either. It is probably that rule that makes an effective jumping robot a fantasy instead of a reality. If you build a bot that cannot damage the arena floor, you probably can't damage your opponent. Assuming your bot could damage your opponent without damaging the floor when it missed, you would most likely never hit your opponent anyway (**Figure 2.4**).

Assuming you could build a control system that allowed you to aim at a moving target, you would still have to aim it while on the floor. Once it is in the air, powered flight is banned; you could not adjust your course to hit an opponent that changed direction after you pressed the jump button. All but the very

Figure 2.4

A jumping bot missing its target.

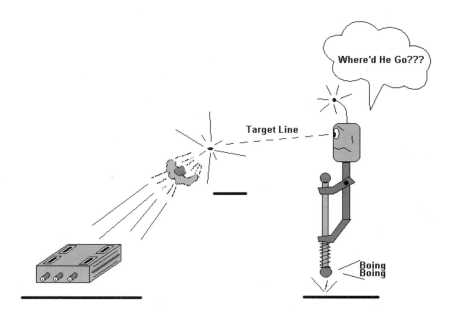

slowest bots can change direction in a split second. It would take your bot longer than that to jump up 6 feet and come down where you thought the opponent was heading. Yes, a jumping attack would look really cool but you'd be lucky to hit anything. Swinging an overhead hammer or spike is your best bet at striking your opponent and doing some damage from above (**Figure 2.5**).

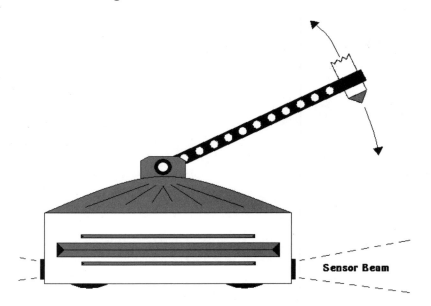

Figure 2.5

An overhead swing attack mechanism.

Builders use different methods for powering the overhead attacks. I've seen just as many powered by electric motors as powered by CO_2. I don't recall seeing any that utilized an ICE to power a fast overhead attack. However, there are some effective bots that use a hydraulically powered mechanism to push a spike or other pointed object into the opponent. **Figure 2.6** shows Lawrence Feir's Brick Slayer, a hydraulic powered, overhead-attack robot.

Stationary Spikes, Wedges, Bumpers

Stationary spikes, wedges, and bumpers (**Figure 2.7**) are the most common weapons among all fighting robots. They may

Figure 2.6

Lawrence Feir's Brick Slayer (Photo courtesy of Lawrence Feir and Team Innovation Robotics)

be considered secondary in nature but their usefulness is undeniable in both offensive and defensive terms. In defensive terms, spikes, wedges, and bumpers can all be considered armor. Use any means to keep your opponent away from the

Figure 2.7

Examples of spikes and wedges.

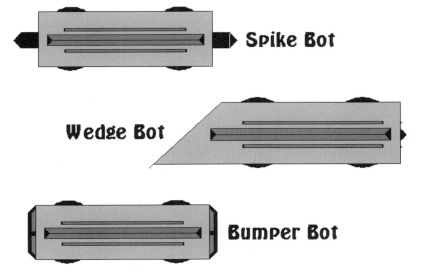

insides of your bot. Any of these three items can keep your opponent at least a safer distance away than if it were right next to you. In offensive terms, the wedge is the most utilized weapon. You see wedges everywhere and on almost every bot. Wedges take the form of skirts around bots like Jason Bardis' Dr. Inferno Jr. The back end of OverKill is a wedge that Christian Carlberg uses to great advantage. Some bots simply ARE the wedge as in the case of Turtle Road Kill by Sam Steyer and Ed Taylor.

Projectiles

Projectiles are allowed in some major competitions. The only restriction is that the projectile itself must be tethered to the robot (**Figure 2.8**). If you fire your bot's projectile and miss the opponent, the projectile must be held by the tether at a certain distance. I think you can see why you can't put a .357 magnum on a bot. It is the same reason you see very few robots that employ projectile weapons. They just aren't that powerful yet, though I have seen one robot team making strides toward a very effective projectile weapon.

Clamping and Smothering Weapons

Derek Young's Complete Control is the first robot effectively to pick up an opponent and give it a body slam. This type of bot

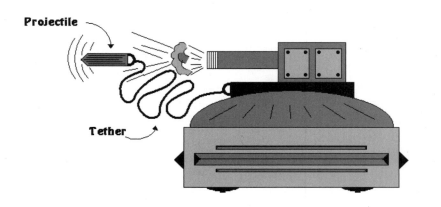

Figure 2.8

A bot firing a projectile weapon that is tethered.

Figure 2.9

Drawing of
clamper-type bot.

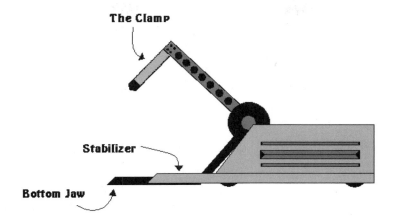

The Clamp

Stabilizer

Bottom Jaw

is called a clamper bot (**Figure 2.9**). There have also been robots that cover up their opponents and try to drag them to the arena hazards. There is really only one special rule that affects these types of bots. The "No Pin" rule states that you can hold your opponent for up to 30 seconds before you have to let go. This rule also keeps you from slamming your opponent into a wall and holding him there until the end of the match.

What's Left?

What is left is up to you and your imagination. Combination bots like spinners with wedges, spinners with lifters, and wedges with lifters are very popular (**Figure 2.10**). Taking the best of two worlds can be an extreme design and construction challenge, especially for the beginning builder.

Multibots are a breed that has been around for a while. Multibot competitors typically build two or three bots that, when combined, make up the total weight limit for a particular weight class. See **Figure 2.11**. They fight as a team against a common opponent unless they happen to be pitted against another multibot team. There are rules that allow you to defeat a multibot without destroying every part of the team. Typically, if 50 percent or more of a multibot has been disabled, it loses. This is why most multibots are made up of at

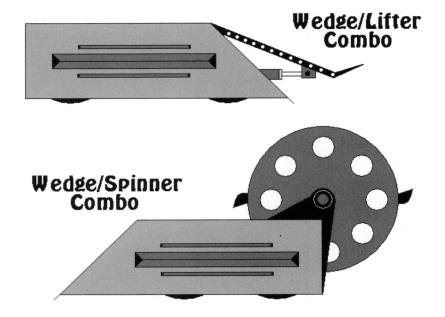

Figure 2.10

Drawing of wedge-lifter and wedge-spinner combo bots.

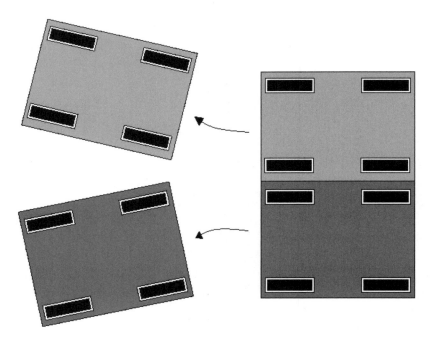

Figure 2.11

One example of a multibot.

least three individual machines. That way you have to disable two out of three to win.

Summary

There are so many things that you might hope to use as a weapon that I just can't list all the rules that allow or forbid them. Read the rules of the competition you plan on attending and follow them. If your weapon isn't covered somehow (not likely), send an e-mail to the event coordinator for a specific opinion.

The next part of the book will cover some of the construction methods used to build the weapons we've talked about. At the end of each new section we'll see how that weapon stacks up against the others. Watch along the way for input from veteran builders.

3

Getting Started

Getting started on your perfect destruction machine can be a daunting task. Balancing what you dream up with what you can accomplish can be very hard to come to terms with and even harder to do. Use your imagination and make sketches of your proposed bot based on what you want it to do. Do you want to pick up your opponent and slam it to the floor? Do you want to rip pieces off your opponent until it is a jerking mass of motors, wires, and ball bearings? Whatever it is that you decide upon, draw it out so that you can start deciding how it will be built and powered.

Second Step

After you've figured out how you want to destroy your opponent, you need to find an opponent. In other words, which competition will you attend? As I mentioned earlier, each competition has its own set of rules. Find them and learn them. If you have questions about them, ask the event organizer. If your weapon of choice is against the rules, it is for good reason. Usually safety is that reason, so the organizer is not likely to change the rule on your account.

There are lots of things to consider when designing your robot. Competition rules are the first. Your budget and available time are the second. Do not bite off more than you can chew in either category. Your knowledge of several subjects is the third. Robotics in any form is a mixture of electronics, mechanics and software. Because of the popularity of the sport, the electronics and software have pretty much been boiled down to buying off-the-shelf parts at specialty robot parts stores on the Internet. The mechanics are where your creativity will shine.

Try to break down your design into the following categories:

1. Frame and structural support.

2. Drive train.

3. Electronics and remote controls.

4. Armor.

5. Weapons.

This book is concerned with weapons only. My other book, *Combat Robots Complete,* will get you up to speed on the other categories. Perhaps the most important thing to remember when getting started is actually to get started. It does not matter how old you are. In fact, if you are old enough to read and understand this book, you are old enough to get started.

Start Young!

I've been in robotics since I was 12 (14 now!), but I am still reluctant to call myself a "veteran." Maybe not a veteran, but I gave myself enough of a head start to be experienced; I had time to get together the trials and errors needed to learn about robotics.

Being young means I have lots of time after school to devote myself to robotics, but before I began, I devoted myself to researching for hours. In robotics, research is

very cumulative, so those facts and numbers I originally learned were the fundamentals of what I know today.

However, being young can be tough, too. It tends to steam me up when someone walks up to my dad to ask about my robot. Or when I've done "pretty well for someone my age." Everyone is pretty understanding once I explain it to them; however, that doesn't solve the problem of not being able to drive to events! And those parental permission forms!

Regardless, I recommend to anyone young who has an interest in robotics to start immediately, and get your "period of adjustment" out of the way.

Will Tatman, Builder of Unlicensed Chiropractor

Figure 3.1

Unlicensed Chiropractor. (Photo courtesy of Will Tatman and Team In-Theory)

Rock–Paper–Scissors

Did you ever play "Rock–Paper–Scissors" as a kid? If so, you know that the rock destroys the scissors, the scissors cut the paper, and the paper covers the rock. No matter what, no one has a real advantage, but each one does well against a certain opponent. Robotic weapon systems are pretty much the same; but in a match, advantage can come from several factors.

Assuming a true rock–paper–scissors type of match between two robots, the real advantage is going to come from a sturdy design, strong construction, and good driving. That is what you need to think about when reading the rest of this book.

Summary

This chapter was a simplified overview of how you should get started in combat robotics. The key to success is to do lots of research and a good bit of design before spending any money. Also, the best weapon you can get is driving practice. Remember the rock–paper–scissors aspect of weapons when designing the killer bot.

The next chapter will talk about building the most used, perhaps the most effective and maybe the most boring weapon in combat robotics: the lowly wedge.

4

The Wedge and the Rammer

The lowly, boring, humble wedge (**Figure 4.1**) is the topic of this chapter. Many builders and spectators of combat robots feel that the wedge is all of those things and less. However, there are not many weapons that can be as easily included in almost any robot design you might think up. For that reason, the simple wedge is also the most used weapon across the field of fighting robots. The wedge is used as the sole weapon of choice, or as a major backup weapon, in some form, on most of the effective robots you see today.

The offensive idea behind the wedge is to give your bot the ability to get under the opponent. The defensive idea behind the wedge is to keep your opponent from getting under you. This makes it easier to push the opponent around and control the match. If the competition arena has hazards, the wedge bot uses those to effect damage to his opponent. Of course, simply slamming the other guy into the wall or wall barrier can eventually disable him. In this respect, the wedge is like a rammer bot but with the advantage of breaking the opponent's traction a little more easily. That is why I'm including the rammer, **Figure 4.2**, in this discussion.

The rammer bot operates on basically the same principles as the wedge bot. Power in the drive train is a big factor in its

Figure 4.1

Drawing of a
simple wedge
bot.

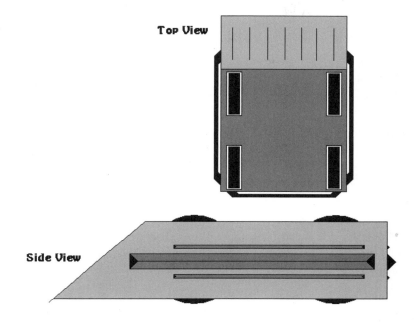

Figure 4.2

Drawing of a
simple rammer
bot.

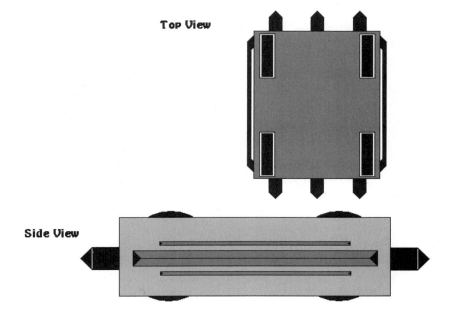

success in the arena. Traction is another big factor. Both the wedge and the rammer can and should be built with heavy materials. Use every bit of weight allowance to get the advantage. Remember that force = mass × velocity. In other words, the heavier it is and the faster it travels, the harder it hits. The rammer needs this hard-hitting power to break the traction of the other bot and push it into arena hazards or the wall. The wedge uses the power for the same reasons but, unlike the rammer, the wedge can put the opponent on its side or upside down. The key design component of wedges and rammers is strength.

Strong and Simple

We started in the mid 1990s, competing in a sport of our own creation called RCombat. Combatants modified small radio-controlled cars and power tools into small combat robots, complete with flame-throwers, rockets, drills, cutting torches, and chainsaws. Basically, the only real rule was "no metal armor," since it was considered an impediment to the event goal of mass carnage and destruction.

All of our robots follow a few key "commandments": invertability, redundancy, robustness, serious traction, and avoid single-point-failure modes. We went through many growing pains in the development and early competition phases working on transmitting significant power to the ground. We kept breaking things to find the next weakest link—there is always room for improvement. Avoid single point failure modes like one radio receiver, one battery circuit, etc. Redundancy will help you stay alive long enough to limp to a judge's decision. One key aspect of robot building and wrangling: don't be nice. Don't hesitate to shove the thing out of the back of a pickup truck or throw concrete blocks onto it. No matter how much you abuse your robot in testing, the damage will pale in comparison to the first 30 seconds of combat.

The right tool for the job. The right material for the application. There seems to be a magical aura around exotic materials. Titanium and magnesium and ceramics OH MY! The reality is, one should use the right material for the application instead of picking the one with the coolest reputation. We use titanium alloys where we need strength and resilience (side armor). We use aluminum where we need light weight (wheels, structure). We use exotic variants of stainless steel when we need strength (drive train). We use brass when we need compliance. There is no reason to build a combat robot entirely out of titanium when aluminum or steel would suffice. The right material for the job is the one that gets you through combat with a victory, and you won't know which one that is until you get in the ring.

The first step to winning a battle is avoiding losing to yourself. Practice driving. Smack your robot into curbs and walls to find your weak shock points. Test drive with all armor installed so you can find out how bad it interferes

Figure 4.3

JuggerBot.
(Courtesy of
Mike Morrow and
Team JuggerBot)

with your radio range. Maintain a "pre-fight" checklist to make sure you don't lose to stupid mistakes. Anytime you see a team JuggerBot member with a clipboard, you see our current fight coordinator. They have only one job— make sure the pre and post fight checklists are followed exactly. Things they check include battery charged, power connections made, key bolts tightened, plugs seated properly (receivers, speed controllers, etc.), tires inflated, weapons armed, etc.

Mike Morrow, Captain of Team JuggerBot
(www.juggerbot.com)

Construction Theory

Again, the wedge and the rammer both depend on strong motors, a strong drive train, and wheels that get lots of traction on the arena floor. Your robot will have to be able to withstand the forces generated when hitting your opponent, forces that you are depending upon to disable that opponent. If your bot cannot take the abuse, you will not win.

The shell of the rammer or wedge should be mounted so that impacts do not transfer the shock load to the frame and drive train. That feat is nearly impossible, so you should concentrate on minimizing the shock load as much as you can. The electronics and batteries should be mounted to the frame so that they don't experience any shock load that happens to get transferred from the shell. Impact surfaces should be reinforced with strong materials. Come to think of it, all of these statements hold true for all types of robots. They are just especially true for the rammer and the wedge, and extra scrutiny should be applied when designing them.

Construction

Because both types of bots are always slamming into their opponents, both the wedge and the rammer can make use of the same construction techniques. Simple, flat-walled boxes

Figure 4.4

Spikes on a
wedge bot.

can represent rammer bots, but in truth they usually employ some type of inactive weapon.

Spikes (**Figure 4.4**) are the most frequent choice, but I have seen people attach saw blades or other sharp objects to the rammer. They can penetrate very light armor but they usually only cause scratches or maybe dents at best. However, scratches and dents are visible to the match judges and can get you some points. Spikes and other sharp edges tend to give you a way to grab onto your opponent and possibly push it around in the direction of your choosing. .

Wedge robots come in a couple flavors. One type of wedge is simply designed to get under your opponent and break his traction. That wedge is usually low, short, and flat. Some builders attach this type of wedge using hinges so that it will move up and down depending on whether or not the bot is upside down. Of course, if you use this method, you should build the bot so that it can run inverted. Dagoth, pictured in **Figure 4.5**, is one such bot. Dagoth also has a spike attached to the front. The wedge is mounted halfway up the rear wall of the bot.

Even if I had not designed the bot to run while inverted, I would still have positioned the wedge in the middle of the rear wall. The wall itself acts as a stop and gives Dagoth the

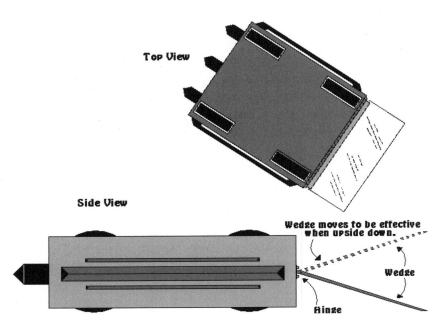

Figure 4.5

Dagoth's wedge.

ability to push his opponent around. If the wedge is mounted at the top of the bot, it will be easy for the opponent to drive right over Dagoth, rendering the wedge useless. Some builders add sturdy hooks or teeth to the top of the bot to keep this from happening.

While I'm talking about the low, flat style of wedge, I should include the parallelogram-type design. This wedge design gets its name from the way it looks. See **Figure 4.6**. This bot can run while inverted and the wedge is effective either way. The rear wedge is not totally useless either. It makes the rear attack approach very difficult for your opponent. Lifters and flippers have to lift higher just to come in contact with you. Spinners may not be adjusted to hit at that height.

Depending on the weight class you are building in, you can make your low, flat wedge out of several types and sizes of materials. In Dagoth's case, a 30-pound bot, I used a single sheet of 1/8-inch thick polycarbonate with absolutely no support structure. Because the polycarbonate is so strong and the bots are so light, the wedge needs no support. This is not the

Figure 4.6

Parallelogram-
style wedge bot.

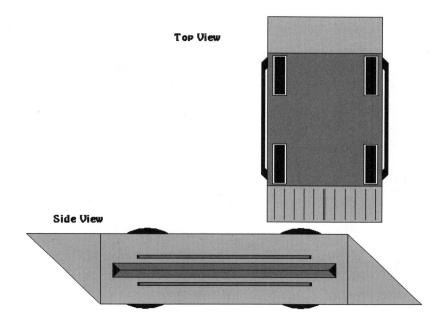

case in the heavier weight classes or even with the parallelogram-type wedges.

Another type of wedge is used to do the same thing as the low,
flat wedge, only it also gives you the opportunity, with enough
strength and speed, to flip your opponent on its side or completely over. That wedge starts out low and slopes up to a
steep incline like a bulldozer blade, such as the one on
MiniRip in **Figure 4.7**.

Wedges should be made of strong material and securely fas-
tened to the bot's frame. Ideally, there should be two frames
for a wedge or rammer bot. The external frame is the part that
contacts the opponent and takes the brunt of all the forces
transferred. Lots of your weight allowance should be placed in
the external frame. Armor should be heavy so that it can withstand hits from spinners. The internal frame can be made of
light materials such as aluminum, polycarbonate, carbon
fiber, and other plastics. As I've said before, the external frame
should be mounted to the internal frame so that impact forces
are minimized on the internal frame. In short, you need to

Figure 4.7

MiniRip's wedge.

shock mount the two frames. Shock mounts can be found in most industrial catalogs or in junkyard treasures. They are usually meant to reduce vibrations between two pieces of equipment. Most of the time, a shock mount is made of rubber with a bolt or stud sticking out. When the rubber base, which could be surrounded by a mounting bracket, is mounted to one frame, the stud or bolt is used to attach the other frame. Use this device to mount electronics and battery compartments to the internal frame as well. Pay attention to shock loads listed in the catalogs, and don't buy them if they are too light-duty.

Rock–Paper–Scissors

Because rammer and wedge bots have strong frames and internal parts that are resistant to shock loads, they are particularly useful against spinner-type bots. They should be built to withstand multiple hits to and from the dangerous spinner weapons. The best strategy is to ram into the spinner and slow its weapon down. If the spinner is not the full-bodied variety, you may be able to get behind it and use a wedge to put the spinner weapon into the floor of the arena, stopping it.

Continue ramming throughout the match and attempt to push it into the arena hazards.

The most important thing you can have to help win any match using a wedge or rammer bot is driving ability. Stay away from the lifting arms of the lifter bot and the clamping mechanism of the clamp bot. Stay out of range of pneumatic spikes and overhead hammer/spike attacks. Aggressiveness and lots of hits on your opponent will get you points and possibly a judge's decision. Because of the caliber of fighting robots in the ring today, it will be very difficult to win a match by knockout. So, get in there and slam your opponent. Try to control the match as much as possible by pushing your opponent around the ring into hazards or the wall.

Summary

Rammer bots are the foundation of fighting robots. A simple, yet strong drive train, a sturdy outer body, and simple inactive weapons make up the basics of any robot that gets in the ring. The wedge adds a bit more offensive strategy by allowing you to get under your opponent and break its traction more easily. Internal and external frames are recommended to reduce shock to the electronics and drive train. Driving practice is the most important part of your wedge or rammer driver's attributes.

In the next chapter, we'll see what goes into building a lifter. We'll discuss the advantages and disadvantages of using pneumatics, electricity, or kinetic energy to power a lifter. Also, we'll hear from a veteran builder of pneumatic robot weaponry.

5

The Lifter and Flipper Bots

The lifter and flipper bots are currently very active in robot combat. From the beginning, it was recognized that many robots could not operate when upside down. That fact ripped an opening in the tournament bracket fabric for the lifter and flipper bot. A fast and easy win could be effected and lifters became very popular very quickly. Because of that, many bot builders started building the self-righting mechanism or including the ability to run inverted. Nevertheless, some of the most successful and entertaining bots are of the lifter or flipper category.

The one big difference between a flipper and a lifter is speed. A flipper needs to actuate very quickly in order to get the opponent completely off the ground and upside down. A lifter can rarely stay in contact with the opponent long enough to accomplish this. With either type of bot, leverage is the key to being successful. Your lifter/flipper arm must be strong enough to do the job both in lifting ability and in structural integrity. Good design practice will take care of the structural part. It's really your choice on what to use to power the lifter/flipper. Hydraulics are traditionally slow and strong. So, they are more suited to lifting. Electric linear actuators are in the same category. Pneumatics are traditionally fast moving,

so they are more suited for use in flippers but can also be used to power a lifter. A friend of mine, and long-time pneumatics user, uses CO_2 to power most of his weapon systems. Though the system is usually some type of hammer or overhead spike attack, Terry's logic is an excellent example of designing for the rules of the competition that you plan to attend.

No matter which power plant you use, lifter and flipper designs sometime have one other advantage. If you design correctly you can include the ability to self-right—something that is helpful when your bot is pitted against another lifter/flipper bot in competition.

Construction Theory

As I've said, the key to successful lifter and flipper bots is leverage. Lifters get the opponent off its wheels. Flippers get the opponent off the ground. Both use some type of arm to get under the opponent. To better understand the concept of a leverage arm, think of the teeter-totter example in **Figure 5.1**.

Figure 5.1

Teeter-totter example.

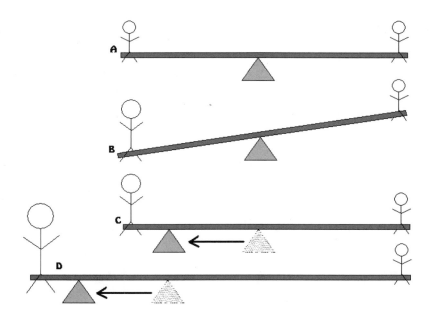

Children of equal weight can sit on each end and balance. Replace one child with an adult and that end sinks to the ground. Move the fulcrum closer to the adult and you will eventually achieve balance again. If you increase the distance between the child and the fulcrum, you must either increase the weight of the adult or decrease the distance between the adult and the fulcrum. This is a very important concept that works the same way when lifting an opponent.

Figure 5.2 shows one type of lifter robot lifting a rammer bot. There are three versions of the lifter in that figure. Version A has the lifter arm mounted to the front of the bot. Notice that version A is tilted off the ground. This is because the builder forgot about leverage. In this case, it is much like the child versus adult teeter-totter version B of **Figure 5.1**. Although the bots weigh the same, the leverage point isn't in the right spot. There are a few different ways to fix this problem. In version B of **Figure 5.2**, you see that the lifter mounting point has been moved toward the center of the lifter bot. This changes the leverage point enough to lift the opponent, solving the tilt problem. This also requires you to get even closer to the opponent to get under him. In version C of **Figure 5.2**, we've added stabilizer bars to the front of the lifter bot. This effectively moves the lifter arm mounting

Figure 5.2

Lifter bot versus rammer bot.

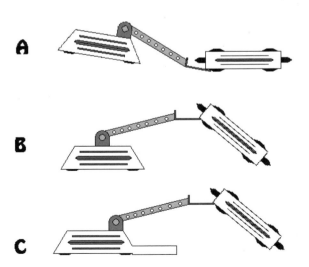

point to the middle of the bot, while keeping the same attack distance as version A.

A gear motor sometimes powers the type of lifter mechanism we've looked at so far. However, a regular gear motor isn't usually strong enough and may need to be geared down further than normal. This, of course, will make it slower than normal. Electric winch motors, used on pickup trucks, have been used successfully as the power source for several lifter bots because the heavy-duty gearing is built in at the factory. This type of power source has an advantage over some of the other types of lifter power sources. Because the motor and shaft can be modeled as a straight shaft that turns a full 360 degrees, as shown in **Figure 5.3**, the lifter bar itself can turn a full 360 degrees if constructed correctly. The advantage is that the lifter bot can use its lifter arm to turn itself over if it happens to get inverted.

Electric linear actuators are gear motors with a special set of gears to change rotational energy into linear energy. In other words, they move in and out instead of in a circle. A linear actuator acts in much the same way as a standard pneumatic or hydraulic cylinder. The main difference is that the pneu-

Figure 5.3

Model of a winch/lifter arm combo.

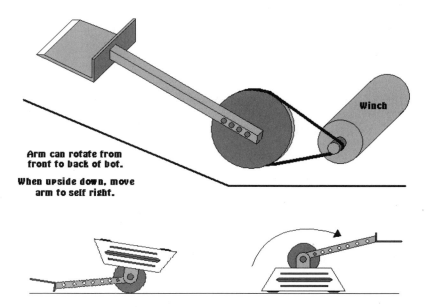

Arm can rotate from
front to back of bot.

When upside down, move
arm to self right.

matic cylinder is much faster. The hydraulic cylinder speed is dependent on the amount of fluid the pump can provide and is slow compared to the pneumatic type. This means that a flipper bot will do best when a pneumatic cylinder is employed. Electric and hydraulic systems are best used for the lifter robot.

A few robots, including the very successful Biohazard from BattleBots, use a special lifting mechanism called a four-bar linkage. **Figure 5.4** shows how a four-bar mechanism is used in a lifter bot. Notice that the lifting "spatula" moves outward as well as upward. This is an advantage over the "straight-up" type of lifter. The spatula is pushed into the opponent while it lifts, earning a better chance at staying under the opponent and maybe even flipping it over. The four-bar mechanism allows one more advantage. It can be folded down nearly flat, certainly flat enough to make a very low-profile robot. This, and great driving, is what has made Biohazard so successful.

Figure 5.5 shows an effective flipper bot design. The flipper bot will most likely use some form of pneumatics as a power source. However, a very high rate of flow is necessary to get the lifter arm moving fast enough and strong enough to bring an opponent off its wheels and into the air. Even though a regular lifter rarely inflicts damage on its opponent, a flipper can

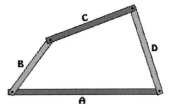

Figure 5.4

Four bar linkage.

Apply pressure to B and C will move up and out.

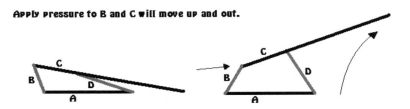

Figure 5.5

Flipper-type bot.

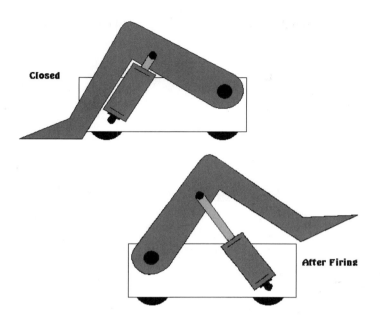

Closed

After Firing

cause great amounts of damage. Some flipper bots can hurl their opponents 6 feet in the air. The crash from hitting the ground can be devastating.

When developing a pneumatic system to power a lifter/flipper bot, you must keep a couple of things in mind. You need a lot of force to lift or flip an opponent. Force equals pressure times the area of the cylinder. Obviously there are two ways to accomplish a specific force. Some bots use large-bore valves, hoses, cylinders, and a big buffer tank to get a very high flow rate in a low-pressure system. The high flow rate is necessary to launch your opponent. If you don't have a high flow rate, you are simply designing a very strong, yet slow, lifter bot. Other bots use high-flow, high-pressure systems that take a lot of knowledge, effort, and money to develop. The flipper bot is typically built from stronger components than the average lifter bot. Thicker materials and more of them are the order of the day. This has to be, since the forces generated within the bot are so much greater than those generated by a simple lifter bot.

Using Pneumatics

In about 1996, I saw the Atlanta Robot Battles at Dragon Con for the first time. I was instantly hooked and began losing sleep pondering my first evil invention to compete. It took me until 1999 to actually complete my first bot, Roadkill 1A, and it certainly lived up to its name. Although my first competitive entry was a dismal failure, it taught me several important things and ensured that robotic combat was in my blood for good.

When I started in robot combat, it was to compete in the Atlanta Robot Battles. This competition is unlike most conventional robotic combat competitions in that there is no arena, so the limitations on destructive weapons are much stricter. Pneumatic hammers (or axes) were the one way I could cause real damage to the other components without the risk of sending shrapnel into the audience. I developed, through trial and error and much studying, a firm grasp of the principles of pneumatics and also a thorough enjoyment in working with pneumatic systems.

Figure 5.6

The Ventilator. (Courtesy of Terry Talton and Team Desade)

Since then, I've installed pneumatic systems in my BattleBots Middleweight, Evil Con Carne, and a lightweight veteran of several local competitions, the Ventilator. I am always amazed by the impressive power of pneumatic systems and am continuously developing more powerful systems. The power to penetrate any opponent's armor is my goal and is getting closer with each new revision of my systems.

Terry Talton, Team Desade (www.teamdesade.com)

Rock–Paper–Scissors

A lifter bot's strategy is the same as a flipper bot's strategy up to a certain point. Both bots require great driving ability. Both require good timing to hit the button that actuates the lifting/flipping arm. After all, it doesn't do any good to hit the button if your arm isn't under your opponent. The difference in strategy comes immediately after pushing the button. The lifter will still need to drive the opponent to an arena hazard, slam it into the wall or, if possible, turn it upside down. The flipper simply needs to line up for another shot once the opponent hits the ground.

Lifters and flippers do well against each other, and the winner is usually the one who spent the most time practicing driving. Driving practice is essential to win against a wedge or rammer bot too. You have to stay away from the wedge part while getting in close enough to get your lifter/flipper arm in a position to be used. Lifters and flippers should have an advantage over the wedge, but only if you design so that the arm will turn your bot over if it gets inverted. This ability should equal your chances against another flipper or lifter if they happen to get a good strike against you first.

The lifter and flipper bot can be effective against a really mean spinner too. However, you have to strike when the time is right. If the spinner is truly mean, you won't have time to run across the arena and attack before it gets its spinning weapon

up to speed. In that case, your bot had better be built to take some horrendous blows. The best thing to do is keep the lifter/flipper arm away from the spinner weapon until you can get it to stop or at least slow down drastically. Once you get the weapon slowed or stopped, attack over and over until the bot breaks. The thwack-type spinner bot uses its drive train to power its spinning weapon, usually built into the frame. This type of spinner can be a problem. It should be really strong and can usually run inverted. So, your bot will have to take some punishment while you slow the attack enough to mount an attack. This can be tricky, since every hit your opponent gets in will count as points. A true knockout on your behalf may be required to win over the many points a regular or a thwack spinner can rack up.

There are a few other types of bots you might face. The overhead attack bots, the pincer and clamper bots, the bots with projectiles, and the multibots are all through the tournament brackets at nearly every event you will attend. Just like the flipper/lifter bots, the overhead attack, the projectiles, and the pincers and clampers all require intense driving and aiming practice to be effective. So, it will boil down to who is the better driver and whose bot can take the punishment.

Multibots can pose a problem though. They are made up of two or more separately controlled opponents that should attack as a team. Each part of a multibot should be smaller and faster than a single, larger opponent, making it more difficult to catch one and lift or flip it. Also, one of them can be attacking your bot while you are distracted by another. The biggest advantage you have is your bot's weight and your driving skill. Pick out the weakest link in their team early on and take it out. Catch one opponent and disable it. Then concentrate on the other part(s) of the multibot. The best thing to say is that you shouldn't have to disable every piece of your opponent. Many events have rules that say you only have to disable a little over 50 percent, by weight, of your multibot opponent. So, another strategy would be to pick out the heaviest member of the opponent team and take it out first. The only problem with

this is that you can be fooled by size in determining which part weighs the most.

Summary

Summing up, the lifter bot came about when someone realized that most of the bots in action could not run while inverted. The creation of the lifter bot brought about the wide-scale ability to self-right or at least to run effectively upside down. The flipper bot was created by enterprising builders who wanted to cause more damage to their opponents while in the ring. Both the lifters and flippers take lots of driving practice to be effective. Most people use electric motors to power their lifter bots. Most people use pneumatics to power their flipper bots. You have to choose between a high-pressure system and a low-pressure system with high volume to power your flipper design. Whether you're building a lifter or a flipper, the design must be structurally strong enough to handle lots of stress caused both by your opponent and yourself.

In the next chapter, we discuss the bots that use an overhead attack. Hammers, spikes, axes, and other objects are put to use in this bot class. This type of bot was designed to take advantage of weak armor on the top of an opponent.

6
Overhead Attacks

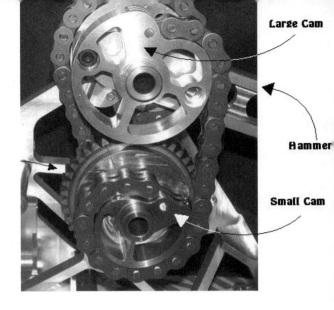

Large Cam

Hammer

Small Cam

This chapter covers the overhead attack weapons such as hammers, spikes, maces, axes, chisels, and whatever else you can think of to swing over the top of the bot to hit your opponent. I've even see a builder use a meat tenderizer. For simplicity's sake, I'll refer to this class of weapon as a hammer weapon. The whole point is to use something to attack the top of the other robot or its wheels. Before the hammer weapon was popular, most bots simply used heavy armor on the sides and thin armor on the tops. In fact, many bots used very thin armor on the bottom until certain televised events started using arena hazards that inflicted damage from beneath.

Hammer weapons have two great advantages if designed and executed correctly. First, they can mount several attacks in a short period of time. Second, the arm can be used to self-right if your bot happens to get inverted. The one thing you need to be wary of is the fact that attacking uses energy, and it is possible to run out of that energy before the match ends. This is most evident in pneumatically powered hammer weapons, but it also shows up in electrically powered ones as well.

Construction Theory

In the building of a hammer weapon, the lever concept pops up once again. This time it is a little different, so we'll look at it a little more closely. **Figure 6.1** shows the three classes of a lever. The first lever is best described by the teeter-totter example in Chapter 5. The fulcrum is somewhere between the load and the applied force. The second lever places the load somewhere between the fulcrum and the applied force. The third lever places the applied force somewhere between the fulcrum and the load. This is the class that most closely represents the hammer weapon.

Looking at **Figure 6.2** you can see why the third class of lever represents the hammer weapon. Simplify the gearing and motor down to a bare shaft and replace the fulcrum with it. The force is still applied somewhere between the fulcrum and the load. How close it is to the fulcrum doesn't really matter. The load is the hammer head itself.

Figure 6.1

Three classes of the lever.

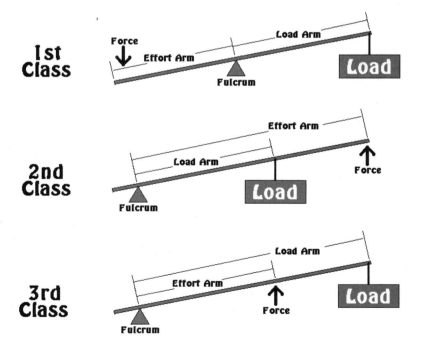

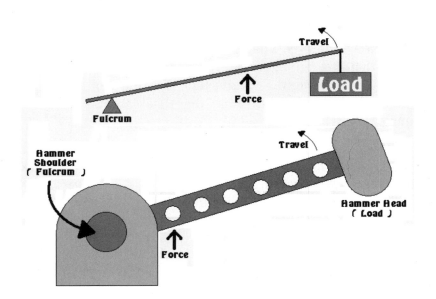

Figure 6.2

Third class lever and hammer arm model.

Figure 6.3 shows another version of the third-class lever. This time a counterweight is added. The counterweight helps balance the load across the fulcrum so that less force is required to swing the load. Of course, if you use the same amount of force to swing the load, using a counterweight will help it get to top speed at a faster rate.

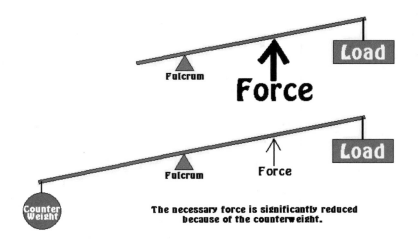

Figure 6.3

Third-class lever with a counter weight.

The faster you swing the hammer, the more damage it does. This is because of the age-old equation: Force = Mass × Acceleration. Force is what does the damage to your opponent. If there is any way you can increase force, do it. Obviously, increasing the weight (mass) of the hammer head will also increase the force. Remember that weight is not really equal to mass.

NOTE: *Weight is a function of mass and the effect of gravity upon it. Mass is the actual amount of matter. The further away from the source of gravity you get, the less an object will weigh. But the mass stays the same. Since I'm not writing a physics textbook, I interchange the terms mass and weight freely. Don't do this in a science class.*

Watch out when increasing the weight of the hammer head. The more it weighs, the slower it swings. So you should find the strongest motor that will comfortably fit in the bot specifications mechanically and electrically. This is exactly where a counterweight can come in handy.

There are several ways to swing the hammer physically. First, the standard pneumatic system can be put to use. As you already know, an electric linear actuator can easily replace the pneumatic actuator. However, swing speed is a key component of causing damage and the electric linear actuator isn't a great choice for the hammer attack. **Figure 6.4** shows a standard way of attaching the pneumatic actuator to a hammer attack weapon. Since a pneumatic actuator is typically very fast, it is probably the best choice. However, there are advantages to other types of power sources that may make them more attractive.

There is a weakness in the way the pneumatic actuator is mounted in **Figure 6.4**. The actuator itself is exposed to attack from your opponent. **Figure 6.5** shows a rack and pinion design powered by the same pneumatic actuator. A rack is basically a gear that has been filleted and laid out flat. The pinion is, of course, the standard gear. The rack and pinion combination changes linear energy into rotational energy, or

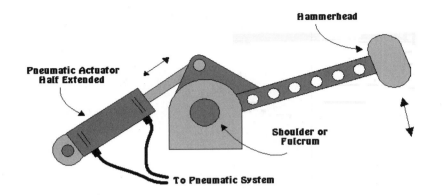

Figure 6.4

Pneumatic actuator powering a hammer weapon.

vice versa, depending on how you want to use them. In this case the pneumatic actuator is pushing and pulling the rack.

The hammer arm is connected to the pinion and rotates throughout the entire pull of the actuator. If you don't want the hammer arm to swing 360 degrees, you will need to specify the size of the gear so that it is twice as big around as the length of the actuator's stroke. The rack should be the same length as, or a little longer than, the actuator's stroke. This will give you a full 180-degree swing. Of course, the main advantage to this kind of setup is the fact that the actuator sits flat within the bot. That presents the opportunity to cover it with armor so that an opponent can't easily damage it.

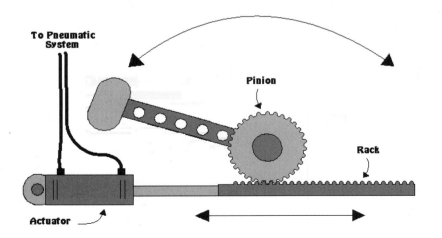

Figure 6.5

Rack and pinion hammer design.

Figure 6.6

Using a gear
motor to power a
hammer weapon.

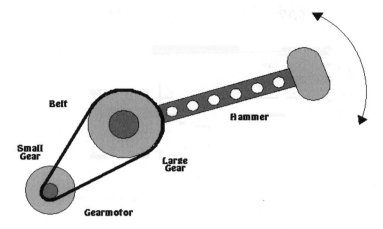

Hammer can travel 360° if mounted correctly.

You can also use a standard gear motor, as shown in **Figure 6.6**, to power the hammer weapon. However, unlike the version in the lifting arm examples of the previous chapter, you need speed to be effective. So the electric winch motor would not be a very good candidate. A very powerful motor that spins very fast will be ideal. Be sure to gear it so that you balance the speed with the weight. This type of power source has the advantage of being able to rotate 360 degrees and can be used to self-right if your bot gets inverted, as long as the force is sufficient.

There is at least one team of builders making technical advances in the overhead hammer attack department. Team Hurtz, www.teamhurtz.com, uses a fusee or snail cam-type reduction to get the most power and speed out of their hammer weapon in their bot called "Beta." Fusee may or may not be a fitting name for this, but it is close. Webster's defines fusee as a conical, spirally grooved pulley in a timepiece from which a cord or chain unwinds onto a barrel containing the spring; by its increasing diameter it compensates for the lessening power of the spring. Looking at **Figure 6.7** you see that the snail shape starts small and gets larger as you travel around its circumference. This cam can't have teeth like a gear because there is no way to keep a second gear meshed with it. However, Team Hurtz uses a chain. It is possible to use cables.

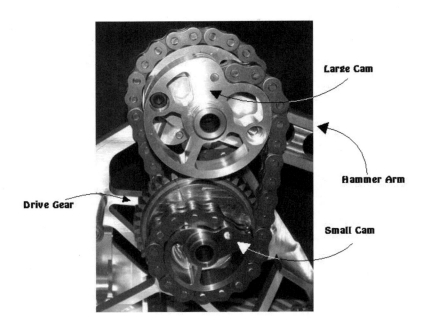

Figure 6.7

Fusee (or snail cam) reduction. (Courtesy of John Reid, Team Hurtz)

Either way, the cam is the fulcrum of the hammer arm. As the chain wraps around the cam, the gearing ratio gets bigger. So, at the start, the cam has a small radius, applying maximum torque to swing the hammer. As the hammer travels, the radius grows, giving more speed at less torque to the hammer arm. This also helps keep the motor running at the optimal speed throughout the entire swing.

One more way to power a hammer weapon is shown in **Figure 6.8**. The spring-loaded hammer can cause significant damage to an opponent. However, spring weapons are notoriously slow at reloading. Typically the builder uses some type of winch or other high-torque, slow-speed method to rewind a cable or chain to pull back the hammer arm. This is a significant drawback, since the biggest advantage to a hammer bot is being able to attack multiple times in a very short period. Even though matches have been won with spring-loaded hammer weapons, I wouldn't recommend them unless you can design the ultimate, high-speed reloader.

Figure 6.9 shows several types of hammer heads that may be used on one of these bots. Blunt objects like a sledge hammer

Figure 6.8

Spring loading a hammer.

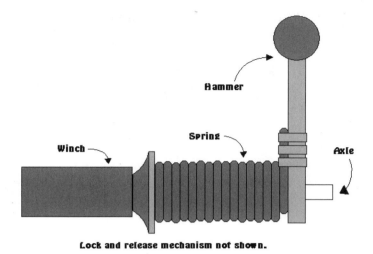

Figure 6.9

Several hammer head types.

head, meat tenderizer head, or a mace cause lots of dents in thin armor, vibrations that could rattle loose a key connector, and can possibly inflict blows that bend frame members, causing misalignment of drive train gears or sprockets. Sharp objects can penetrate thick armor if applied with enough force. Blows from sharp objects could possibly penetrate sen-

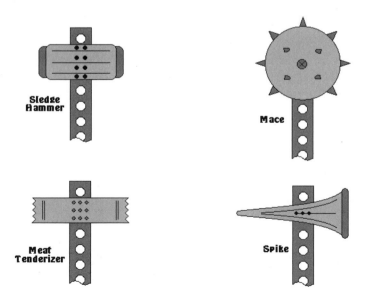

sitive control electronics. Tires can be punctured and gears can be damaged so that they no longer mesh correctly.

You've heard of the thwack bot, the robot that attacks by spinning horizontally on its axis. **Figure 6.10** shows a couple types of overhead thwack bot designs. This one is more like an overhead hammer attack, but it still uses the drive train to create the attacking force. These bots usually have two-wheel drive, making them harder to drive and aim. The attack is initiated by reversing direction rapidly. The hammer has to be light enough for the drive train to power its attack while at the same time being heavy enough to actually do some damage. Most thwack bot designs keep the entire bot body within the diameter of the drive wheels. This makes the hammer section easier to flip when reversing directions. However, at least one bot, by Christian Carlburg of Team Cool Robots (www.coolrobots.com), doesn't follow this philosophy. Christian's bot, called OverKill, has very large tires, a very large knife-like attack arm and a secondary wedge like the top example in **Figure 6.10**. Despite the disadvantages of the bot's being harder to drive and aim, Christian has won many times with OverKill. It all comes down to driving practice.

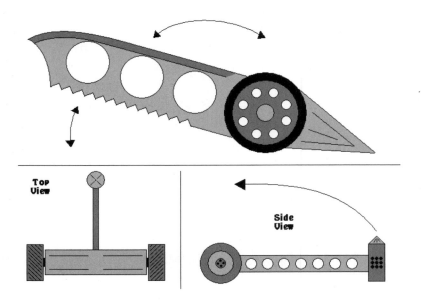

Figure 6.10

Overhead thwack bot designs.

There has been some speculation that a hollow, liquid-filled hammer head might do more damage than a solid head. I don't believe anyone did any math to confirm or deny the theory. I won't either. If you want to experiment to find out the answer, then go for it. Just remember that the hollow head has to be big enough to hold a significant amount of liquid. Also, the liquid should probably be denser than regular water. Mercury might be a possible candidate, but it is a poison and is probably against the rules of whatever event you plan to attend. It is my opinion that your time is better spent concentrating on perfecting your drive train, armor, and driving ability than trying to develop a liquid-filled hammer head.

Critical Hits

How do you destroy your opposing robot? Take out their wheels? But what if it's tracked? Flip them over, but what if they are invertible? No, a critical hit would be best—one that would disable an aggressor permanently without possibility of parole. A metal spike through a radio receiver, speed controller, or any number of items internal to the speed bump in your road to glory.

This is the thought pattern that generated the overhead swinging weapon design of the super heavy weight Isosceles. The point behind the spike weapon being that, once an enemy's electronic innards have been pierced, you will obtain sudden victory. The problem lies in getting there.

With the electronic piercing theory, the spiked weapon is obvious, but how is it best implemented? It could be mounted horizontally (such as Shaft) and rammed. However, most bots are well armored from the front and back, and the sides usually contain the drive media, such as tires or tracks, effectively armoring our fragile target. The spike could be swung for greater effect, such as Gold Digger, however the effect is still not the sudden victory that a pierced, smoldering battery pack would generate.

For Isosceles, the choice was elementary. The spike had to come from overhead. Most bots are not as well armored from the top. Many also have large flat top surfaces that not only are useless at deflecting blows, but also invite well placed overhead swinging spikes in for a bite of lunch. Due to weight class restrictions, most bots are built with little space between the electronics they house and the housing itself. This is the weakness that the overhead swinging spike takes advantage of.

To power the weapon only two sources were considered— pneumatics and electric. Electric power was chosen due to the seemingly limitless number of hits that could be produced and the frequency with which they could be delivered. This was also in keeping with Isosceles' strategy of speed and getting in, delivering hits, and getting out before the oppositions' weapon could come to bear.

Included in Isosceles' design was a defense to the same weapon that we chose for offense. We called it the Doghouse design. The delicate electronics were housed in an elongated steel teepee that resembled a rolling dog-

Figure 6.11

Isosceles. (Courtesy of Eric Koss and Team Triborg)

house. Isosceles' weapons, speed, and defense were proven in battle against No Apologies, a flat topped, slow mover with an overhead swinging weapon. Isosceles used its speed advantage to get around the weapons of No Apologies, deliver successive hits (with its faster electrically powered spike), exit the danger zone, and reposition for another strike. Throughout the two-minute match, Isosceles received only two hits from the stronger weapon of No Apologies, which were of no consequence due to the defensive doghouse design. The quick overhead weapon of Isosceles disabled the stronger bot's weapon with its repeated hits, but alas, fell in battle due to an inferior speed controller cooling design.

Eric Koss, Team Triborg (www.kosscomputers.com)

Rock–Paper–Scissors

The main strategy of any overhead hammer attack is to rack up a lot of points by inflicting several hits. This can accomplished through driving and aiming practice. Hammer attack bots have an advantage over wedge and rammer bots in that they can score more hits more often. The rammer or wedge bot needs to back away from its opponent, and charge again to score another point. The hammer attack bot can stay in one place, right next to its opponent and still collect lots of points with multiple hits. Lifters and flippers suffer a small disadvantage against a hammer attack bot. The lifter and flipper attack modes pretty much depend on getting the opponent upside down. The typical overhead attack bot has the ability to self-right and continue its attack. Hence the disadvantage to the lifters and flippers. When it comes to spinners, I still have to give them the advantage. They cause much more damage in a fight and can get in bigger blows. I shouldn't even mention the possibility of getting the hammer weapon ripped off while they are attacking a spinner. Pincer and clamper bots can grab the hammer bot, but I see the contest as a tie, since the hammer will most likely still be able to attack while captured. Hammer bots make great opponents against a multibots. The

ability to strike in two places almost at once can be a major advantage against two or more opponents.

Summary

As you can see, success depends on the strength of your machine and the amount of driving practice you subject yourself to. After that, it boils down to the luck of the draw in the tournament bracket. You can use whichever type of hammer head you like, depending on what you want to accomplish. Just remember that speed and weight affect your striking ability drastically.

In the next chapter, we'll discuss a few types of fighting robots: the crusher, hugger, and clamper bots. They are similar enough to be considered at the same time. The clamper usually tries to get one grabber arm under the opponent and one on top, and then carry its opponent to an arena hazard. Occasionally the opponent can be picked up and body-slammed. The hugger is the horizontal form of this attack, without the body slam. The crusher is strong enough to defeat its opponent by squeezing the life from it once it is caught. Either way, it's a fairly interesting method of attack.

7

Crushers, Huggers, and Clampers

This chapter is really about three types of robots that have a similar mode of attack. The crusher bot is probably the most dangerous since it should be able to destroy its opponent simply by closing its jaws. This type of bot was made famous by a British bot called Razer (www.razer.co.uk). The hugger bot usually attacks by wrapping a set of arms around its opponent and carrying it to one or more arena hazards. It should be possible to build a hugger bot that is strong enough to crush an opponent but I haven't seen it yet. The clamper bot, made famous by Complete Control, a bot by Automatum Technologies (www.automatum.com), grabs the opponent at the bottom and the top and picks it up in the air.

Construction Theory

When building a hugger bot, you need to pay attention to the hugging device. I've seen a couple of methods of construction but so far I haven't seen the one method that would probably make the hugger bot a success.

One way to build the hug device is to use a winch or other high-torque, low-speed electric actuator to push or pull the horizontal jaws closed. **Figure 7.1** shows this method. The

Figure 7.1

Winch and linear actuator powered hug device.

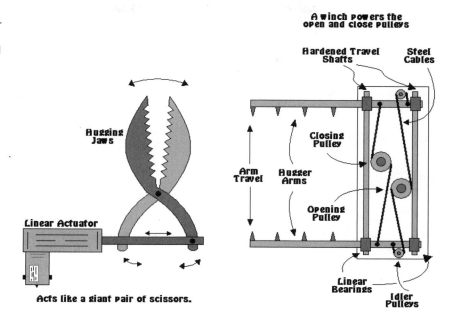

A winch powers the open and close pulleys

Hardened Travel Shafts

Steel Cables

Hugging Jaws

Closing Pulley

Arm Travel

Hugger Arms

Opening Pulley

Linear Actuator

Linear Bearings

Idler Pulleys

Acts like a giant pair of scissors.

winch method is a little more difficult to build, not to mention more unreliable. That's because you have to concentrate on keeping the cables or chains tight throughout the entire opening and closing process. Of course, if your opponent hits the bot and the frame flexes at all, your cables or chains will most likely get out of line and screw up your entire method of attack. Using an electric linear actuator will solve this problem. However, both methods are slow acting. This will give your opponent time to make several attacks while you are trying to get it in your bot's jaws.

A pneumatic actuator can replace a linear actuator with a little bit of work. This will solve the problem of a slow attack but it does have its own disadvantage when using the typical CO_2 system. Pneumatics act like a spring. CO_2 (referred to simply as gas from here on) is typically stored under pressure so that you can get more attacks out of a smaller, denser supply tank. A little bit is let out of the supply per attack. When it is let out into your weapon system, the gas expands, pushing open any pneumatic actuators on the system. The drawback lies in the fact that if you apply enough pressure on the actuator, you can

recompress the gas in the cylinder and open it. This means an opponent might be able to wiggle out of your grasp. You might be able to get around this problem by using gasses that aren't as easily compressed, but the components of those systems are typically larger and heavier.

Another way to escape the compression problem is to employ something that does not compress at all. Hydraulic systems have this advantage. Liquid can be considered a solid when confined to an enclosed system. Take a log splitter as an example. You could use gas to power a log splitter. If you did, the splitting awl would slam into the log end and stop. Pressure would continue to build until there was enough to split the log. At that point, the awl would slam through the log very quickly and then slam into the stop at the end of the splitter. This would be very dangerous, because a piece of the log could fly off the splitter, not to mention the possibility of losing pieces of the splitter itself. Since liquid can't be compressed, the awl will travel at a constant speed throughout the travel space. As long as the pump pressure is greater than

Figure 7.2

A typical crusher bot design. (Courtesy of Feir and Team Innovation Robotics)

what is required to split the log, the awl will not stop. This same idea can be applied to a hugger bot. The main drawback is the fact that hydraulic systems are heavy and can be difficult to employ in a light-weight bot.

Crusher bots typically focus all their power into a small point. Some builders use a chisel for the point. Some use a spike. Some don't do anything special at all, using the design of the crushing arm to inflict damage. Either way, the main idea is to get as much crushing force as possible into the smallest amount of space.

Since you won't get the amount of force needed to crush an opponent with pneumatics or electric motors, hydraulics is the way to go when building a crusher. However, you have to design carefully to fit the extra weight of the heavy hydraulic components into the weight limit of your robot. I have seen some very compact hydraulic systems but they are fairly expensive. Even if you get a small, powerful system, you still have to design your bot so that it can withstand the amount of force you'll be applying. At several events, I've seen bots that successfully house a very powerful hydraulic system that fails when trying to crush an opponent. The most common failure point is the crushing arm itself. If not designed or built correctly, the crushing arm will buckle and bend instead of harming an opponent. The second most common failure point is the joint where the crushing arm meets the frame of the robot. The frame should be constructed to handle the force as well.

Hydraulic systems are normally slow. As you know, slow attacks are not the best attacks. There are special hydraulic systems that pump the fluid into the cylinder at high speed with little force until resistance is met. Then the pump switches over to low-speed, high-force mode. With a variable displacement pump, your crushing arm can close quickly until it meets the armor of your opponent and then start the slow squeeze.

Overall, a well-designed crushing weapon can decisively destroy your opponent bot if you can catch it. Your best bet is

to leave a bit of wiggle room in your weight limit so that you can include a quick, drivable frame. Then get some driving practice.

Figure 7.3 shows the components of a typical clamp bot. The job of this type of bot is to get under and over the opponent. Then, if the clamp is strong enough, pick up the opponent and carry it to an arena hazard. Occasionally, you might get lucky enough to get a lift that is positioned right and be able to drop your opponent. This might cause damage, but it's more likely to earn points instead.

The typical clamp bot is basically a lifter bot with an extra arm. The lifter section should follow all the recommendations in Chapter 5. Leverage is key, but the ability to move your opponent around the arena is equally important. The only exception is that the clamp bot must be able to carry an opponent's entire weight around the arena.

The clamp bot's extra arm is used to apply pressure to the top of the opponent. The two actions create a clamping force. Once the opponent is clamped, the lifter arm can pick it up. The clamp arm is not meant to do damage. If it were strong enough

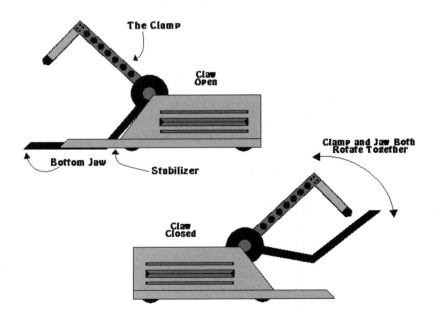

Figure 7.3

Typical clamp bot design.

to do damage, it would probably be too weak to hold its opponent. The clamp arm should be able to close quickly in order to catch a fast opponent. A fast linear actuator is probably the best thing to use, but a gear motor has been used as well.

Leverage is more important in the clamp bot than it is in either the crusher or the hugger bots. Since you need to be able to get above the tallest robot with the clamp arm, it is likely that your clamp arm/lifter arm pair will be on the front of your bot to get the greatest possible reach. Just as in a regular lifter bot, the back end will tilt up instead of lifting your opponent. To compensate for this, you should design some front-end leverage points, as in **Figure 7.4**. It wouldn't hurt to place your motors, batteries, and whatever else is fairly heavy toward the rear end of your bot.

Figure 7.4

Clamp bot with and without leverage.

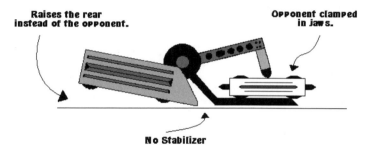

Raises the rear instead of the opponent.

Opponent clamped in jaws.

No Stabilizer

Radicus

I suppose I went about designing my first bot bass-ackwards in that I had a strange desire to make it a cube. Because I realized I had no real engineering skills it seemed to me a cube would be easy enough to make, and that saw blades would be both fun and easy to implement as weapons. In designing the bot I realized pretty early on that I couldn't exactly have one set of wheels that stuck out both the top and the bottom of the bot at the same time—due to the simple geometry of a cube—so instead I elected to have a "top drive" and "bottom drive" to compensate on the off chance I was flipped. The front and rear weapons were a given, seeing that the robot's name was "Palindrome."

It was easy to reconcile the choice between doing extra work of building a second drive train and being invertible, or not doing the extra work and potentially being knocked out if I was flipped over. Sure—it made Palindrome 2.0 a very compact bot, but that wasn't really a problem for me. The problem lay in trying to figure out how to attach the top and the bottom drive trains to the central power distribution block without losing fingers from getting them caught in between the drive trains and the central body.

For Palindrome 3, I've reluctantly moved away from the cube-shaped body idea. Although it was a very distinctive design, I've since actually won a match or two with some of my other 'bots, and I'd like Palindrome to be a little more competitive next time. This meant a trade-off in body size to reduce the center of gravity, although Palindrome 3 will have tracks, so it'll be just as cool!

I started out by wondering "Could I do that?" after catching an episode of BattleBots one night. Two short months later Palindrome 1 died a death in its first match at a BattleBots competition. That was it for me—I had the bug! I think it's very important to ask yourself why you're doing this, because otherwise the mounting expenses and the negativity of potentially losing many matches before winning anything can really take a toll if you're not careful. I lost 9 straight matches with various bots before finally winning a match a year and a half later. The real reason for the drought is because I try to build fairly ambitious bots, shufflers, tracked bots, bots with saw blades for weapons. Common sense and history seem to have shown us that none of those styles of design are usually winners, but I want to challenge those theories, and I believe that any bot, no matter what the design, can prove to be the "rock" to some other bot's "scissors." You just have to watch out for the "paper"... For me, the losing streak wasn't a big deal though. The enjoyment of designing, the accomplishment in building, and the exhilaration

Figure 7.5

Palindrome 3.
(Courtesy of Tony
Hall and Team
Radicus)

of competing, are plenty enough reward for what is after all a bloody expensive hobby!

Tony Hall, Team Radicus (www.radicus.net)

Rock–Paper–Scissors

There are two main drawbacks to these three types of bots. One is that your bot must get in really close to its opponent to actually do some damage. The second is the complexity of the mechanisms. Most bots have to get in close, but these three have to stay in constant contact with their opponents to inflict damage. As for the second drawback, well, everyone knows that simple is better. There's a well-known fundamental engineering rule called KISS (Keep It Simple Stupid). These three bots almost fly in the face of that rule. However, you can build simple mechanisms and join them together for success.

The ultimate enemy of either of these bots is the full-body spinner. This type of spinner attacks from all sides and can be very difficult to stop and grab. In fact, with any spinner, there is a high likelihood that the clamp arm, the lift arm, or both

will be ripped off if you attempt an attack before the spinner is stopped. If your bot is in the ring with a lifter or flipper, the winner will probably be determined by the driver with the most skills. A clamper, crusher, or hugger can disable hammer bots. The main thing you should do is try to attack on the sides, instead of directly in the line of fire of the hammer. The hammer should be able to disable the more complex mechanisms of these three bots if given a chance. Wedges and rammers are the most susceptible to a clamper, hugger or crusher attack since they depend on their wheels being on the ground at all times. Multibots can cause a significant problem. Your mode of attack is typically slow. While you are concentrating on one target, the other(s) will be attacking your bot, gaining points.

Summary

This chapter covered three bots that I consider to be in the same class. The clamper, hugger, and crusher all have similar modes of attack as well as similar strengths and weaknesses. The hugger bot typically uses some method to close a pair of arms around an opponent and then carry it to an arena hazard to inflict damage. The crusher bot typically employs a powerful hydraulic system and a pointed attack arm to concentrate the most crushing force into the smallest amount of space. Once an opponent is caught, the idea is to pierce through armor to get to the delicate insides or at least to bend the frame enough to disable it. The clamper bot takes the design of a lifter one step forward in the line of evolution. With its added arm, and a good bit of complexity, the clamper grabs its opponent, picks it up and carries it to an arena hazard or drops it for the body slam. Once again, driving practice is going to be the key to success with either of these three bots.

The next chapter will cover one of the most dreamed-about, yet one of the least-implemented weapon systems in combat robotics. Projectile weapons are subject to several restrictive rules that make it very difficult to build an effective system. However, there are ways to do so.

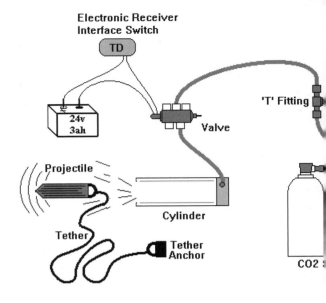

8

Projectile Weapons

The first time someone is asked to come up with a weapon for a combat robot, the answer is usually to strap some kind of shotgun to it. The only problem with that is the fact that it is banned by at least three different rules. There is even a rule that makes the use of a legal projectile weapon fairly difficult. One rule says you can't have explosives. Another rule says you can't have chemical reactions. That is sort of a redundant rule because the explosion is considered a chemical reaction. The third rule, and the one that truly hinders the building of a real projectile weapon, is the one stating that all projectiles have to be tethered to the bot. That means the part that is actually being thrown at the opponent must be secured to your bot. Usually there is a limit to the distance that the projectile is allowed to travel. Once you have a legal projectile weapon, you have to decide whether or not it is a one-time-use weapon or if you want to reload. Either way, you should try to make sure the tether isn't considered an entanglement weapon. Some events do allow entanglement weapons, but most do not.

Construction Theory

Some form of pneumatics has powered every legal projectile weapon I've seen. In fact, I can't imagine any other power

source being used. Then again, I'm sure there are some people out there with a bigger creative side than I have. If you are that person, feel free to let me know about it *after you build it.*

Figure 8.1 shows a tethered projectile design that should fit within the rules. Again, you still need to check with the event organizer to see whether or not the tether could be considered an entanglement device once the projectile has been fired.

Figure 8.2 shows a standard pneumatic system design. You should notice that **Figures 8.1** and **8.2** are almost exactly the same. There are only three differences. The end of the pneumatic cylinder has been removed. The plunger has been changed to a tethered projectile. And finally, the return section of the gas line has been eliminated. The reason for this is that when you design a pneumatically powered projectile system, you are actually designing a contained pneumatic system failure. The normal pneumatic system keeps the plunger within the cylinder. In the failed system the plunger exits the far end of the cylinder and inflicts damage on whatever it hits. The big difference is the fact that the plunger (projectile) is tied to the robot.

Regular pneumatic systems are dangerous to mess with as designed. This makes a projectile system even more danger-

Figure 8.1

Potential tethered projectile design.

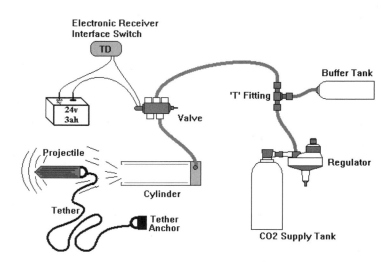

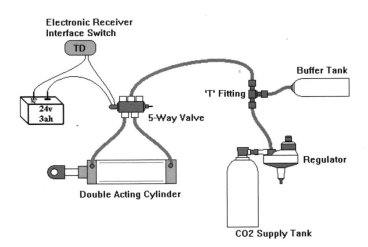

Figure 8.2

Standard
pneumatic
system design.

ous. Never stand in front of the projectile even when you think there is no gas in the system. Never aim the weapon at anything other than the intended target. You should also develop a system for firing your projectile weapon when testing it. A checklist, including every item of the pneumatic system, should be drawn up. Check it before firing to ensure there are no surprises. Once the checklist is finished, clearly announce to all those around that you will be firing your weapon. Then loudly count down from five to one in order to give any bystanders a chance to stop you. After "one," yell "fire" and trigger your weapon. Always wear safety goggles and stay clear of the end of the barrel. Last, and certainly not least, do not attempt to build any pneumatic system without training or without someone who has experience.

Rock–Paper–Scissors

Building a legal projectile weapon system is difficult enough, but the fact is they are also hard to use effectively. If you design to drop your projectile after firing, you don't have any chance at a second shot. If you design multiple barrels, you still have to be able to hit your opponent. Event arenas to date have a single place for drivers to stand. Your robot can be pointing at any angle, and your opponent can be at any other

angle. Even if you line up where you think you will hit your opponent, you could be off by only a few degrees and miss it completely. Even if you hit your opponent, the chances of a perfect hit against a flat surface are very remote. Even if you find a flat surface and hit it perfectly, 1/4-inch thick polycarbonate has a tensile strength of 9,000 pounds per square inch. Without multiple hits on the same spot, it will be very difficult to do serious damage.

With as many drawbacks the projectile weapon has, I can't really recommend it against any other bot. If the projectile weapon is the only weapon on your bot, the odds are that you won't ever cause any damage to your opponent. With that said, I'd love to see a successful projectile weapon on a bot. Consider this a challenge to prove me wrong.

Summary

The projectile weapon is basically a designed pneumatic system failure. It is not strapping a gun to your bot. You have to pay attention to an event's pneumatics rules, as well as its specific projectile rules and even its entanglement rules. If you build one, make sure you follow the strictest safety precautions when operating it.

In the next chapter we'll talk about multibots. Multibots are two or more machines that make up the total weight allowance in a class. Building a multibot can give you the advantage of multiple attacks and, if done correctly, can make it more difficult for your opponent to win by knockout.

9

Multibots

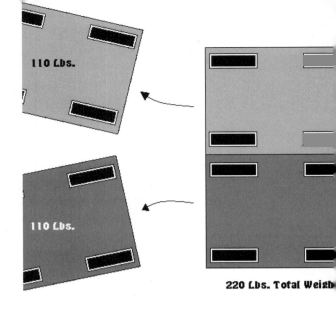

110 Lbs.

110 Lbs.

220 Lbs. Total Weigh

Multibots aren't so much a class as they are a means to a win. A multibot is two or more bots that split a total weight allowance and fight together as a team. Some teams build multibots so that they can divide their opponent's attention between several targets. Some teams build multibots to make it more difficult for their opponent to win. Still other builders build multibots in order to experiment with different weapons systems. The main advantage is that you increase all the advantages of each team member by the number of bots in the team. Of course, the main disadvantage to multibots is that you increase all your problems by the same factor.

Construction Theory

The teams who build multibots as a distraction usually build one big robot with a weapon and two or more small bots that run around but can't cause any damage. The Chiabot from BattleBots comes to mind. This is one large bot with a spinner weapon. Sometime during the match, two small remote control car toys shoot out from within the Chiabot and drive circles around their opponent. Hopefully, the driver of the opponent is distracted enough to let the big bot get in and attack. **Figure 9.1** shows an approximation of the distraction multi-

Figure 9.1

Distraction
multibot.

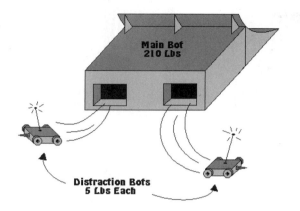

bot, which takes advantage of the multibot class but not the intended benefits of the class.

Figure 9.2 shows another multibot that actually takes advantage of the spirit of the multibot rules. The rules for most events say that 50 percent or more (by weight) of the total weight of the bots must be disabled for the match to be over. You can see that in the case of the distraction type of multibot, all you need to do is disable the biggest section. If you intend to take advantage of the 50 percent rule, you should split your

Figure 9.2

The equal
weights multibot.

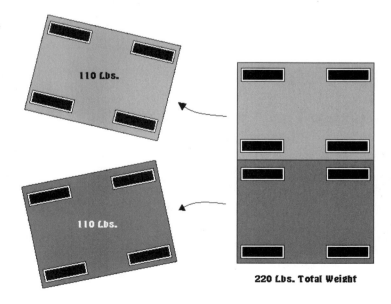

weight allowance between three multibot sections. I wouldn't recommend more than that because your bots will start getting too small to do any damage to their opponent. So, if you have a 120-pound weight limit, each bot should weight about 40 pounds. This way, if one bot gets destroyed, the other two can continue attacking up until a second bot is destroyed. It should be obvious that having one member of your multibot team weigh 50 percent or more is a bad thing.

Probably the most fun and technically beneficial type of multibot is the weapons experiment shown in **Figure 9.3**. This is the type of multibot that makes you learn more, since each individual bot has a different type of weapon. One bot could have a spinning disk. A second bot could have a lifting arm. A third bot could be a big wedge. This bot can also inspire two or more teams to get together and become one team using each individual's specialty. Each team, whether in different states or in the same neighborhood, can take a portion of the weight allowance to build its own new team member. On the other hand, each team can get together and build the other team's specialty using that team for advice. This way, everyone gets to learn something new and build something different.

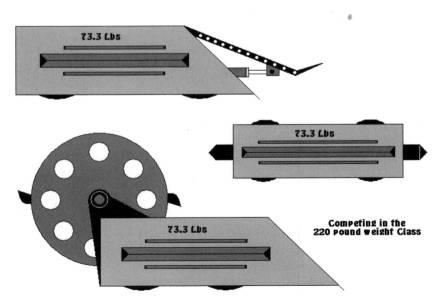

Figure 9.3

The weapons experiment multibot.

Figure 9.4

The swarm theory.

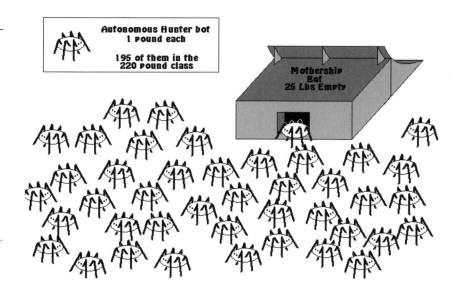

Autonomous Hunter bot
1 Pound each

195 of them in the
220 Pound class

Mothership
Bot
25 Lbs Empty

Figure 9.4 shows what I call the swarm theory. There is one "mothership" bot that contains lots of tiny, autonomous fighting robots. In this swarm, the mothership bot marks the opponent target so that each of the tiny bots can recognize it as the enemy. Once the mark is established, each tiny robot homes in on the enemy and attacks, inflicting lots of tiny injuries that could add up to a win.

So far, the swarm theory is just that, a theory. No one has been able to turn it into a working system for several equally daunting reasons. So we'll probably not see this type of multibot for quite a while. One problem is that it can be very difficult to mark your opponent for the tiny bots to recognize it as an enemy. One method would be to paint your opponent with an infrared laser that the tiny bots will detect. That will prove extremely difficult, since your target is always moving, not to mention trying to attack you. Another method might be to attach a magnet to your opponent. The magnet could house some type of beacon that attracts all the tiny robots. Your main problem will be the fact that not all of your opponents are made of steel. Also, most of your opponents don't use steel as armor. The magnet might just slip off. Another possibility is a

sticker that houses the beacon. Either way, you have to get close to and make contact with your opponent, when your objective is to stay away and let your little guys do the dirty work.

Once a form of beacon is in place or the tiny bots have some other way of recognizing their target, they need weapons. Today's combat robots are built from many types of materials, including titanium, aluminum, steel, and polycarbonate (bullet-resistant glass). These materials can be difficult to damage with one big robot of the same weight. It will prove very difficult to build a weapon that can inflict damage on an opponent that's ten or twenty times its size. Maybe the best course of action would be to send the little fellows out like kamikazes to get tangled up in the gears and wheels of an opponent. The only problem with that is electronics and mechanics that would get wasted on every match. As much as I would enjoy seeing the autonomous swarm attack implemented, I'm afraid that it is pretty far down the road of combat robotics development.

Rock–Paper–Scissors

This is a short chapter mostly because multibots really aren't a class of fighter but an undertaking that includes a few different classes all at once. In fact, a multibot can combine any number of weapon classes. There are limitless possibilities in the combinations. Since that is the case, we really can't figure out what might happen between a multibot and any given opponent unless we know what combination we will build.

If you figure out which weapons your multibot members will employ, you should be able to use the rest of the book to figure out your best mode of attack against any given opponent. Just remember to take into account the fact that your multibot team member will weigh less than its opponent. That might seem a daunting task, but remember that there is strength in numbers. Get lots of driving practice with your teammates. Create attack strategies and practice those. Put it all to use when you get in the ring and you might come out a winner.

Summary

This chapter talked about the multibot, a set of bots that split up a weight class. Doing this for distraction has only one advantage. It's more fun for more people. If you want to be really effective, you need to figure out which combination of weapons and which combination of robot weights will benefit you the most against your opponent. Driving practice and planning play even more of a role with this class than the others if you want to win matches.

The next chapter will discuss one of the oldest types of weapon in history, not to mention combat robotics: spears, spikes, and other types of pokers. We'll look at how they are powered and what type of design is best for achieving different objectives.

10
Spears, Spikes, and Pokers

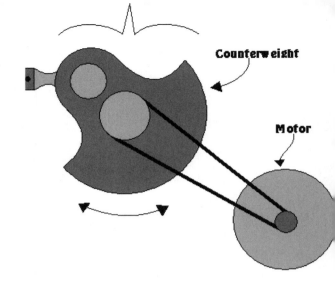

Counterweight

Motor

Spears, spikes, and other types of pokers have nearly the same purpose in robot combat. The main difference is that a spear is usually under some type of power and has movement, while a spike or poker does not. Because of that, I'll really only talk about the spear. The spikes and other pokers have their place mainly as defensive weapons, though some people do use them offensively. They can be put to good use flattening tires and possibly breaking armor. In order to accomplish that task, the spike must be attached to a very strong drive train and a structurally sound frame. That is why so many wedges and rammers use stationary spikes as secondary weapons. The structure and drive train are already there.

Spears, the active ones, need a bot that has plenty of traction. A single powered spear will most likely cause a dent in the opponent's armor and then push it away. You will spend a lot of time getting back within striking distance after a shot. That is, if the spear doesn't bend and jam. Then again, spears do not necessarily have to be used to penetrate an opponent. If you design it correctly you can use a spear along with a wedge to overturn your opponent completely.

Construction Theory

There are currently three realistic ways to power a spear weapon. The most common and effective way is to employ pneumatics. By now you should know that the secret to effective pneumatic weapons is a high flow rate. Large-bore valves, hoses, and cylinders will go a long way but not quite far enough. If you use CO_2, it has to change from liquid to gas within the system. This takes time and energy. Allowing the CO_2 to dump into a buffer tank that has a volume larger than the cylinder can negate the time aspect. Each time the cylinder fires, it fills with CO_2 from the buffer. While the buffer is emptying into the cylinder, the CO_2 tank is refilling the buffer. This is where the energy aspect appears. The CO_2 doesn't magically change from liquid to gas. Pressure on the CO_2 keeps it in a liquid state. Energy in the form of heat changes the liquid to gas when the pressure is decreased from the CO_2 tank, through the regulator, and into the rest of the system. The heat comes from the warm air surrounding the CO_2 tank. All the heat is sucked out of the air and the walls of the tank. The moisture stays and collects on the tank in the form of ice. The ice on the tank robs the entire process of heat, so less and less of the CO_2 changes from liquid to gas. This is the dilemma that comes with using CO_2. The more you fire your weapon, the less powerful it becomes. You can escape this problem by using nitrogen (N_2) or high-pressure air (HPA), but those systems are difficult and expensive to build, not to mention more dangerous. You can't escape the heat problem by attaching any form of heater to the CO_2 tank. For now, that is specifically disallowed in the rules of the major tournaments.

Some people have used large springs to fire a spear. This would work much like a projectile weapon except the spear never leaves the bot. However you still have to reload it. Reloading a spring-loaded spear can be accomplished by using a strong gear motor as a winch. But, just as in the projectile example, this can take a long time. That time is going to allow your opponent to beat on you and possibly win the match. **Figure 10.1** shows an example of a spring-loaded spear weapon.

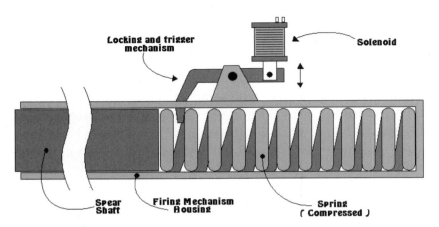

Figure 10.1

Spring loaded weapon.

The next method of powering a spear weapon uses either an electric motor or a gasoline engine. By using a cam or crankshaft mechanism to push the spear forward and return it to the original position, this type of weapon is more like a sewing machine than anything else. **Figure 10.2** shows a spear attached to a crankshaft mechanism.

In general, it will be very difficult to hit your opponent more than once, if at all, at maximum power. Using a pneumatic system for power, the spear is most powerful towards the end of the piston travel. When you are driving your bot, it will be

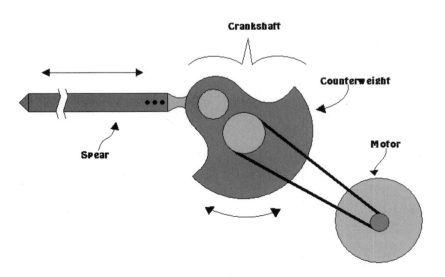

Figure 10.2

Crankshaft operated spear.

extremely difficult to gauge the distance between the bot and your opponent so that the spear will strike at the optimum range. Even when you do, the spear will most likely push your opponent away, and you'll have to do it all over again. Depending on how you construct a spring-loaded spear, the best time to hit your opponent will be somewhere near the middle of the spear's travel. With a cam- or crankshaft-based spear, the most powerful hit will always be in the middle of the stroke, since after the midpoint, the spear will be slowing down in order for the cam or crank to return it to the original position.

You will need to decide which type of spearhead to use on your spear. You can't simply use the end of a pneumatic cylinder arm. You can sharpen the end of a hardened steel or titanium spear. Both of those will be fairly time consuming to sharpen. You can also attach different spearhead styles to your spear. The first head shown in **Figure 10.3** isn't the best head for piercing armor. The rest of the heads are better. A long, thin point will have the greatest success but will also have the highest chance of breaking off. If you do get the spear through the armor of an opponent, then it is likely that the spear will get stuck inside. This is bad, since your weapon will effectively attach your bot to your opponent. Once it starts moving around, there is an excellent chance that the spear will be bent

Figure 10.3

Spearhead designs.

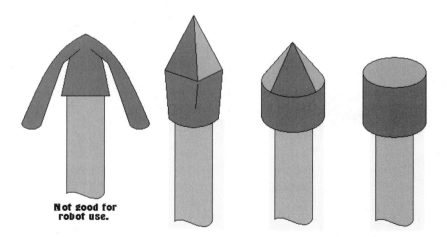

Not good for robot use.

and become useless. If you don't want to take that chance, you can always use a hardened blunt tip on your spear. This tip will only put dents in armor, but the impact might shake loose a key connector or even cause a poorly made switch to turn off.

Rock–Paper–Scissors

The active spear is one of the weapons that I would not recommend. You have to aim it even more carefully than you would aim a spinning weapon. Even if you hit your opponent, the chances of doing damage are very slim, since most bots have heavy armor and may have sloped sides. Even if the side is perpendicular to the floor, your bot must be lined up perpendicular to the opponent to get a good hit. Otherwise, the spear will probably just slide off the initial impact point. Rammers typically have very heavy armor. Wedges definitely have the sloped edges. With either of these opponents, your best bet for success will be to out-drive them. Lifters and flippers can also possess the heavy armor and sloped sides. An extended lifter or flipper arm will, if the lifter or flipper misses, and you get the chance to attack through the opening, form the only possible advantage. Good luck. Overhead hammer bots might give you the same slim advantage. Your spear can damage the more complex workings of crushers, pincers, and clampers, if you can aim and fire it with great accuracy. Multibots pose the least hassle for a spear bot since the parts that create the whole are smaller and weigh less. You may even be able to pierce their thinner armor. Spinners are the most dangerous enemy for a spear bot. The simple act of hitting a spinner can easily bend your spear and turn it into trash.

Summary

If you insist on building a spear bot, you had better get some driving practice. Whether it is powered by pneumatics, a spring, or some type of rotational energy, the chances are that

your spear will get bent and you won't have a weapon. Don't get me wrong. There are some very cool fighting robots with spear weapons. Some of those are successful too. But, it's because they combine a wedge or some other type of secondary weapon with the spear.

The next chapter will be about the most dangerous class of combat robot. This type of weapon has flung more pieces of armor, exploded more gearboxes, and destroyed more entire bots than any other class. It has also been the cause for new safety rules, safety procedures, thicker armor, and sturdier arenas. It also has the greatest number of discernable variations within its class.

11

Spinners

The spinner class is my favorite type of robot. Spinners cause the greatest damage to their opponents. Consequently they are the most-feared types of bot. I've never heard anyone look at the tournament bracket and say "Why did I have to draw this guy in the first round" about a wedge, lifter, or hammer bot. However, I have heard that exact sentiment at every single tournament that I've attended. Invariably the builder is talking about some deadly spinner bot. This is not without reason. As I said before, spinners have caused the most damage, the most innovations, and the most rule changes. Unfortunately, they've caused the most arena breaches and the most injuries. If you are building a spinner weapon, please use the utmost caution.

The spinner class is also the most fragmented of robot classes. There are at least six different types of common spinner weapons. Some of those even have variations. We'll look at each one separately. No matter how many types of spinner bot exist, there is one governing law that dictates whether or not your spinner will be effective. It's the same one governing all the rest of the bot classes. Force = Mass × Acceleration (F = MA). As before, I'll use the terms mass and weight interchangeably because this isn't a physics text. I'll do the same thing with speed and acceleration. All this means is that the force that you

Figure 11.1

F=MA
demonstration
part A.

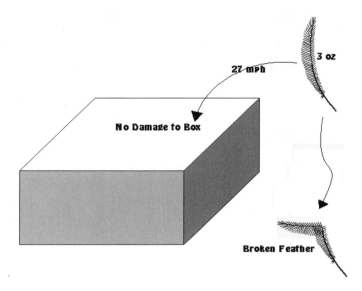

hit your opponent with is dependent upon the weight and speed of the object you are using to hit the opponent.

Figure 11.1 shows a feather hitting a metal box. The feather is traveling at 27 miles per hour and weighs about 3 ounces. The feather breaks when it hits the box. The box is not damaged. I guess the obvious assumption you should make from this is not to use a feather to hit your opponent. However, the one I'm trying to make is that the weight of your striking weapon is important.

In **Figure 11.2** we've exchanged the feather for a hammer that weighs about 10 pounds and travels at about 27 miles per hour. The box is damaged but not beyond repair. The weight of the striking weapon was increased and so was the damage inflicted on the opponent.

In **Figure 11.3** we increase the weight of the hammer to 100 pounds and left the speed at 27 miles per hour. This has caused some real damage to the metal box. In fact, it might be very difficult to repair. Once again, the weight of the weapon was increased and so was the damage.

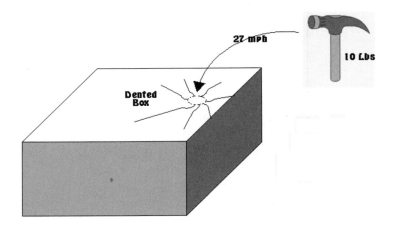

Figure 11.2

F=MA
demonstration
part B.

Figure 11.4 shows what happens on the other side of the spectrum. We've decreased the speed of the attack and kept the same weight. Now the 100-pound hammer is swinging at only 1 mile per hour and there is no damage to the metal box.

After reading all this and looking at the damage done by a very heavy hammer, you might think that the best course of action would be to use an extremely heavy hammer and an extremely high rate of speed. Well, that's true, but it is easier said than done. You do have weight limits to adhere to. This affects the weight of the hammer you swing in two ways. Your hammer can't exceed the weight limit, and you have to be able to swing

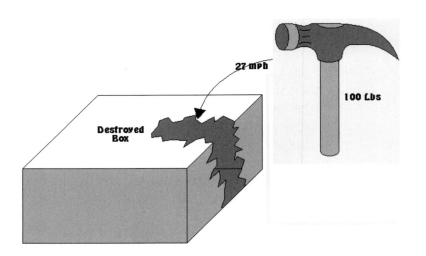

Figure 11.3

F=MA
demonstration
part C.

Figure 11.4

F=MA
demonstration
part D.

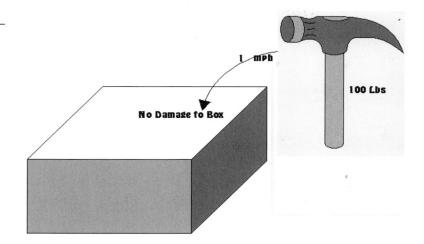

it. The heavier the hammer, the bigger the power source you must use to get it up to a speed that will cause damage. Since the power source of every spinner in existence is either an electric motor or an internal combustion engine (ICE) and not some imaginary, weightless gravity machine, you must include its weight in the limit. This certainly does not mean that you can't build an effective spinner. It simply means you must find the appropriate balance between weapon and power source. Just remember that the heavier your weapon is, the longer it will take to get up to speed. A spinner that takes a long time to reach its top speed is not going to be as effective as one that only takes a few seconds. In fact, that is the very weakness that most opponents take advantage of when fighting a spinner bot.

You can think of a spinning weapon as acting like a spring that is always wound up. Both the spring and the spinner store energy that can be released in an instant. The main difference is that a spring weapon needs to be released mechanically, and it loses energy during its swing. The spinning weapon doesn't require the trigger. Its energy is released when an opponent is hit, and the amount of energy is the same every time, as long as the weight and speed of the spinner don't change.

One other thing that you should know to help your spinner weapon endeavors is the fact that spinning weapons are more

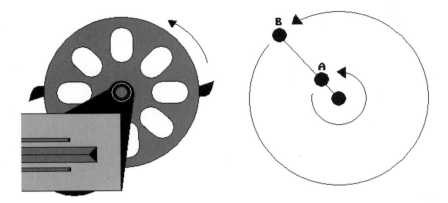

Figure 11.5

Weapon mass placement.

powerful if you concentrate most of their weight around the outside of the spinning mass. **Figure 11.5** shows a spinning disk where most of the weight is around the outside edge. It also shows a diagram that you can use to imagine why this happens; the reasons revolve around the familiar F=MA equation. Points A and B are attached to each other and the center point by a single spoke. Points A and B are revolving around the center point. Imagine that point A can travel the circumference of the circle indicated in 3 seconds. Since points A and B are connected, point B must also travel its circle in 3 seconds. However, point B's circle is quite a bit larger than point A's circle. All this means point B must be traveling faster than point A relative to the center point. If that is the case, it must hold true that all points farther away from the center than point A must be traveling faster than point A.

As you know from **Figure 11.6**, you can model a disk with many points and spokes. If you take all this into account, along with the F=MA equation and what you know about it, it only makes sense that the more points that are traveling faster, the more force your weapon will have.

Construction Theory

As I said, there are several types of spinner weapons. There are the hockey puck spinners, the vertical disks and bars, the horizontal disks and bars, the angled disks and bars, the

Figure 11.6

Modeling a disk with points and spokes.

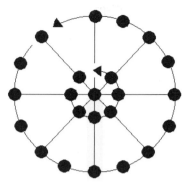

drums, and the saw blades. We're going to take a look at how each type of spinner is constructed and its advantages and disadvantages.

Hockey Puck Spinners

First up is what I call the hockey puck spinner. This spinner resembles a huge hockey puck with teeth. The entire outer shell spins at a high rate of speed. **Figure 11.7** shows a hockey puck spinner called The Revolutionist.

Figure 11.7

Hockey puck spinner, The Revolutionist (Courtesy of Brian Nave and Team Logicom)

Why Spinners Rule and Other Robotic Musings

Combat robotics is all about efficiency. How can you get the most bang for the buck, the most hit for the dollar, the most power for the weight. One design criteria is to build with exotic components and high-tech, high-dollar items. Me, being the cheapskate that I am, chose a different path. Many builders spend their weight allowance armoring their bot and then adding a weapon, for vice-versa. I thought it would be a REALLY good idea if the armor WAS the weapon, efficiency of mass. By completely surrounding the robot with high strength material that is ALSO used as the weapon you can effectively kill two birds with one stone. A full-bodied spinner is just the ticket. Sure, there are other types of spinners that spin some weight around, but there is NO other kind of spinner that can spin the same ratio of weight as a FBS and STILL be supremely well armored.

Another tough task for combat robotics is generating a BIG hit really fast. One of your goals should be to impart as much energy as possible into your opponent as quickly as possible. The faster the energy is added to their system the more violent the absorption must be. Some people use compressed gas, or springs, or ramming speed, but the advantage a spinner has is that it can be storing up kinetic energy the WHOLE time the bots are not in contact. When the bots make contact, all the stored kinetic energy is imparted in BIG hits. Compare that to a compressed air system where the driver maneuvers his bot into striking distance: He will then trigger the weapon, the air will start to expand into the actuator and begin to move it. The energy in the compressed gas is transferred to the actuator in the form of kinetic energy and the moving mass starts to store it up. When the actuator hits the opponent, whatever level of kinetic energy is imparted to the opponent, but unfortunately you are talking about storage times of less than a second. With a FBS the kinetic energy storage times are typically 2–6 seconds. You can store a LOT of power in that amount of time.

Keys to winning with spinning:

- Spin as much mass as FAST as possible.

- Speed is more important than mass past a certain point.

- Energy increases to the square of the speed, and only linearly to the mass.

- Your spinning mass must get to hurting speed in 2 seconds and deadly speed in 4 seconds.

- Make a driveline that resists spinning (i.e., 4WD).

- Use electronic yaw control (i.e., gyroscopic stabilization).

- Make the base and spindle support SUPREMELY stiff and strong.

- For weight efficiency, it is better to have one motor pulling 200 amps than two motors pulling 100 amps. It makes the same torque and weighs half as much. Put your weight in the batteries, not the motors.

- KEEP spinning. With a spinner you MUST spin to win.

Brian Nave, Team Logicom (www.teamlogicom.com)

The Revolutionist's shell spins around its entire drive train, creating not only a formidable weapon but also some very tough armor. The main advantage to this type of system is that you can't touch the bot without taking some type of damage. Spinning steel and ripping teeth protect every direction from an opponent's attack. There are two disadvantages to this system other than spin up time. Probably the more important disadvantage is that with a totally enclosed shell, you can't tell which way the base is pointing. That means you won't know which way forward is.

Builders use a variety of different methods to solve this problem. One of the most prevalent is the addition of a tail or a flag sticking straight up from a stationary position at the center of

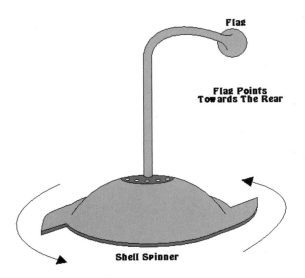

Flag

**Flag Points
Towards The Rear**

Shell Spinner

Figure 11.8

Flag method of direction indication for hockey puck spinners.

the spinning shell. **Figure 11.8** shows this method. The flag pole is bent toward the rear of the drive train and the driver can easily see which way is forward. This is the method used by Ziggo, the famed BattleBots combatant.

Figure 11.9 shows how The Revolutionist remedies this problem. Brian has put holes in the top of the spinning shell and replaced the metal with polycarbonate. Polycarbonate is clear and, in effect, makes a good, strong window to the inside elements of the bot. Inside the bot, on the drive train itself, Brian placed two colored lights. One is on the front and one is on

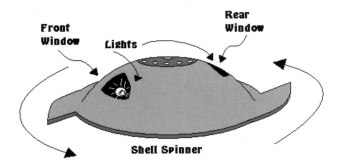

**Front
Window**   **Lights**   **Rear
Window**

Shell Spinner

Lights are stationary on the base.

Figure 11.9

Light method of direction indication for hockey puck spinners.

the rear. As the shell is spinning, the windows blur together to make it look like the entire top is transparent. The lights show the direction of travel, not to mention that they look cool.

The shell also causes the next disadvantage. Whenever there is a mass that is spinning, whatever it is attached to wants to spin as well. The inertia of the spinning shell is felt by the drive train. So, when the shell is spinning and you are trying to drive straight, your bot will actually start to swerve. There are only two ways to fix this. The hard way is to learn to adjust your driving to compensate.

The easy way is to add a gyro. Gyro is short for gyroscope. The one I'm talking about is used in model helicopters. The gyro senses spin direction and compensates by altering your driving commands for you. If you command it to go straight and the bot starts to turn a little to the right, the gyro increases the speed of the right motor in order to straighten out the bot's path.

One other driving difficulty that the hockey puck will experience is the phenomenon of oversteering and understeering. **Figure 11.10** shows what I mean. Suppose your spinner is spinning in a clockwise direction. If you try to turn to the right, the inertia of the shell will tend to make your bot turn a

Figure 11.10

Over- and under-steering.

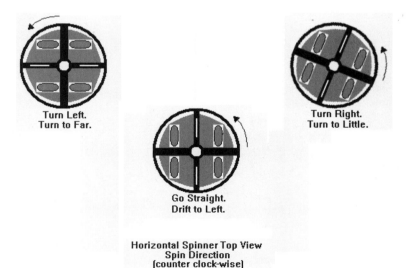

Turn Left.
Turn to Far.

Turn Right.
Turn to Little.

Go Straight.
Drift to Left.

Horizontal Spinner Top View
Spin Direction
(counter clock-wise)

little farther than you would like. If you try to turn to the left, the inertia will work against you and your bot will tend to turn a little less than you want.

Vertical Disks and Bars

I'm going to talk about vertical spinning disks and vertical spinning bars as if they were one in the same weapon since they are basically the same. The main difference and reason for using a spinning bar instead of a disk is the weight limit. A bar is lighter than a disk and can be fitted into a robot that is near the maximum allowed weight limit.

The vertical spinning disk weapon type, shown in **Figure 11.11**, has one main advantage over the hockey puck style of spinner. The main weapon does not take up the entire bot. In other words, you can add a wedge, spikes, or whatever else to your bot and have a combination of attacks or a backup just in case the spinner weapon malfunctions. This add-on ability comes at the fairly high price of having to give up a good bit of striking power, but if you build correctly, it will be possible to strike many times, inflicting the same amount of damage.

The most effective direction of spin for a vertical spinning weapon is from the ground up. Any vertical spinner that spins its weapon down toward the ground and its opponent is taking a big chance. The bot will at least try to climb over the top of

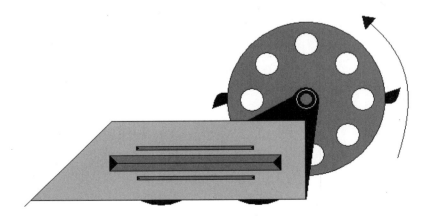

Figure 11.11

Vertical spinner bot.

the opponent, diminishing the damage done by a hit. At worst, a hit could cause the spinner bot to flip over backwards and be disabled. On the other hand, a vertical spinning disk spinning from the ground up has an excellent chance at flipping an opponent over if it doesn't rip off a piece of armor. All the force experienced by the attacking bot is directed toward the ground, unlike the side forces experienced by a hockey puck spinner.

All spinner weapons act like gyroscopes. Whatever direction the weapon is spinning, that is the direction in which it wants to continue to spin. It is effectively the same way the model helicopter gyro works, only this comes naturally instead of being created with electronics. **Figure 11.12** shows the type of forces experienced by a vertical spinner weapon. Since the vertical spinning weapon wants to stay in one position, turning the robot will make it tilt. Be careful that you don't turn too tight a curve or the bot may end up on its side.

Figure 11.12

Gyroscopic forces experienced by a vertical spinner.

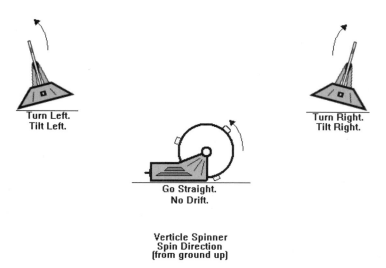

Turn Left.
Tilt Left.

Turn Right.
Tilt Right.

Go Straight.
No Drift.

Verticle Spinner
Spin Direction
(from ground up)

Horizontal Disks and Bars

This next group of spinners has a distinctive attack that has proven to be very effective. Again, I'll refer to both the disks and the bars together.

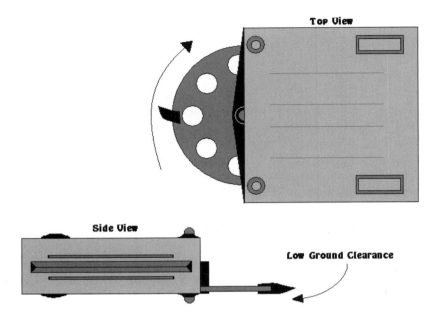

Figure 11.13

A horizontal disk
spinner.

Figure 11.13 shows an example of a horizontal disk spinner. This type of spinner concentrates on attacking the wheels and drive train of its opponent. The best configuration is to have the disk as low to the ground as possible. This not only makes it possible to attack any opponent, no matter how short, but also gives you a shot at the wheels. Everyone knows that if you can't move, you are out of the fight and a wheelless bot doesn't move around very well.

Like all the rest of the spinners, the horizontal disk bot experiences gyroscopic forces. However, it isn't affected as much by the swerving force since most of these bots are four-wheel drive. That means they won't usually require a gyro in order to make a straight run at an opponent. The horizontal disk spinner does feel the same force as a hockey puck spinner when it hits the opponent and can be thrown off course by the impact. Because of this, the driver will probably have to spend a little time getting in line to hit the opponent again. This may or may not be a bad thing. If your weapon takes a few seconds to get back to speed, this may actually be advantageous. This type of bot also experiences the oversteer problem seen by the

117

hockey pucks, shown in **Figure 11.10**, but not to the same extent.

Angled Disks and Bars

Again, I'll refer to the disks and bars as the same thing. This time I'm talking about the disks that spin at an angle relative to the rest of the bot. **Figure 11.14** shows an example of the angled spinner bot.

The angled spinner bot sort of takes the good and the bad from both the horizontal and the vertical spinner bots. Even though one of my most successful bots was an angled spinner called the Six Million Dollar Mouse, I've got no idea why someone would build one other than the reason I had. When the Mouse started out, it was a completely different horizontal spinner bot with an incredible design flaw. The horizontal spinner was about 15 inches off the ground. There wasn't much chance of hitting an opponent and the simplest way to correct that problem was to modify it into an angled spinner.

Figure 11.14

Angled spinner bot.

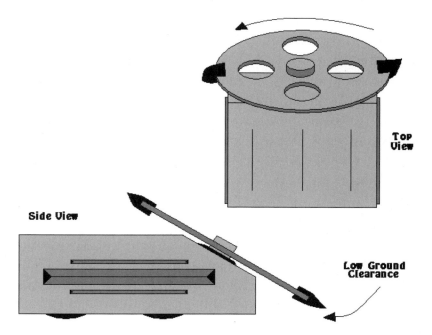

The worst part about an angled spinner is the fact that, depending on your bot's body configuration, there are three attack angles. The main, and the best, attack angle is directly in front, where the spinner is closest to the ground. You can take out an opponent's wheels and armor fairly easily at that point. The other two attack angles are on the sides. One side is okay. The other side is terrible. One side closely resembles the proper spinning direction of a vertical spinner, where the disk is lifting up on an opponent and the hitting force felt by the attacker is directed toward the ground. Unfortunately, the other side of the spinner is traveling the wrong way. If you get into a situation where your opponent attacks your bot on that side, your bot could be overturned.

The Drum

The spinning drum is an innovation that came about through the overall nature of the spinning weapon and the endeavor to overcome a disadvantage of the vertical spinning weapon. **Figure 11.15** shows a typical spinning drum bot.

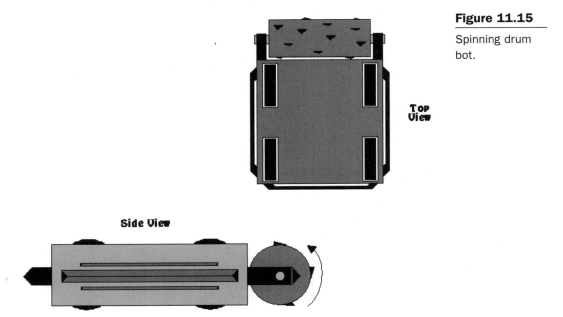

Figure 11.15

Spinning drum bot.

Top View

Side View

You already know that you get more hitting force if you put more weight around the outside edge of a spinning weapon. The drum places most of its weight around the outside edge. One disadvantage of a vertical spinner is the very narrow attacking focus. Often, a vertical spinning disk is only 1/2-inch thick or less. This means you have to get that 1/2-inch lined up with your opponent. With a drum spinner, the drum is typically nearly as wide as the entire bot, making it easier to line up to hit your opponent.

Even though the weight of the spinning drum is spread out across the length of the drum, it still gets the advantage of being on the outside edge. Because the weight is on the outside edge and spread out, you can make your drum diameter smaller than the typical spinner disk. With a smaller diameter, it can be possible to fit the entire drum within the diameter of the drive wheels. This would give you the advantage of invertibility. If you plan on that advantage you had better be able to change the spin direction of the drum so that it will always be spinning upward whether your bot is right side up or inverted.

The spinning drum bot is even more like a standard vertical spinner. There is only one place, even though it is a lot wider now, to attack your opponent. So you must try to keep that drum pointed at the opponent. The flip side of that coin is that you can add a wedge or some other type of secondary weapon to the other side of the bot.

Saw Blades

Saw blades have been used on combat robots since the inception of the sport. However, there has rarely, if ever, been a match where a saw blade-armed bot actually cut its opponent in half. **Figure 11.16** shows a saw blade bot.

There are several types of saw blade to choose from. You can use a metal cutting blade in hopes that your opponent will be made of a compatible metal and will stand still long enough for you to actually do some damage. That's not very likely.

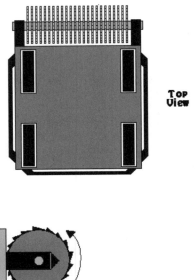

Figure 11.16

A bot armed with a saw blade.

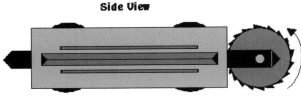

You can choose a wood cutting blade and hope the opponent is made of some type of wood. I've only ever seen one wooden bot, and it got turned into toothpicks within about 10 seconds in its first fight. You can choose a very large wood cutting blade with the aim of ripping off parts of your opponent. That one actually works when the blade is sturdy enough. You can choose a milling cutter and again hope that your opponent stands still long enough for it to be effective. Abrasive cutoff blades work well against metals, but they don't take impacts well at all. They tend to shatter when an opponent slams into them at any angle. The best blade to choose is the emergency rescue blade used by your local rescue squad to cut cars open at the scene of accidents. These blades can cut through nearly any material if they are spinning fast enough and have a strong enough power source.

Other Details

So far I've been combining the spinning bar with the spinning disk. As I said, that's because the bar is just a cross section of the disk and there isn't much difference between the two

when you build. However, the bar is much weaker than the disk. Many bars will bend or even break at the spinning axis. This is not only dangerous, but you will be left without a weapon if it happens.

Figure 11.17 shows a few different spinning mass weapons that have been used in combat. The first is the plain old spinning bar. Steel can be used, but even 3/8-inch steel will bend like clay after a few good hits. Aluminum is out of the question. Not only is it too light, but it is too soft. Stainless steel is a little harder than regular steel, but it won't perform much better. Probably the best material to use would be titanium. That stuff can get a bit expensive though. The second spinner is a weighted bar. This concept follows the rule of thumb telling us to put as much weight as possible on the outside surface. This time the weight is on the outside ends of a thin, hard bar. Again, the bar should be made of titanium if you can get it. The last example is like a two-sided flail. Large, heavy balls, with or without spikes, hang from chains attached to the end of the bar. This really isn't a very good weapon. Each bot that has employed this weapon, that I have observed, has lost one or both of the heavy balls when it hit its opponent.

Figure 11.17

Different spinning mass weapons.

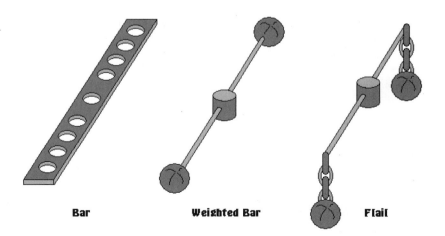

Bar   Weighted Bar   Flail

More on Spinners

When my teammates and I first started building robots, the last thing we thought about was the weapons system—there are enough things to worry about just with getting the chassis to move around! Our first vertical spinner weapon fitted to Kritical Mass was admittedly an afterthought. It was conceived and implemented only days before the robot shipped off to Los Angeles for Robotica I, and it was far from optimum. Our "spinning blade o' death" version 1.0 had many flaws—it was far too thin, it was made of a soft steel alloy, the drive shaft was too small, and it didn't have a clutch. We broke many motor shafts and destroyed many sets of ball bearings due to the shock loads imposed whenever we hit something. The blade also didn't weigh enough, so it wasn't really that effective.

We redesigned our weapon for Manta (spinning blade 'o death version 2.0) using 1" thick hardened S7 tool steel—this took care of any bending issues, and it also weighed more and carried more energy on impact. We also used a 3/4" driveshaft, tapered roller bearings, and a clutch that took care of any significant shock loads imposed on the drive motor. This design was much more effective, winning us several matches in the Robot Wars series—but it still wasn't perfect. The design of the blade was basically a ring with three teeth on it. The ring portion would limit the "depth of cut" of the blade and wouldn't let it deliver a full devastating impact.

With Ultra Violence, we designed our blade once again—this time we were going for something seriously more intimidating. The design of our spinning blade o' death version 3.0 was significantly more aggressive, with an open-tooth style. This would allow the full energy of the 11-pound blade to be delivered to our opponent. We also increased the size of the drive motor and increased the speed (calculated at over 300 mph) in order to increase the amount of energy stored in the spinning blade. This

Figure 11.18

Ultra Violence.
(Courtesy of Jeff
Cesnick and
Team Suspect)

time we had a winner—the blade was by far the most effective one to date. If you're planning on designing and building a weapon like this, please remember to always treat it with respect—they are potentially deadly! Be safe, and happy robot building.

Jeff Cesnik, Team Suspect (www.cesnik.com)

Rock–Paper–Scissors

The hockey puck spinner has a couple of natural enemies. The rammer should be built strong enough to withstand the multiple hits it will take to slow the puck down. If a wedge is strong enough, it could withstand those multiple hits long enough to get under the puck and push it into an arena hazard or even turn it over. Lifters and flippers should be wary of the puck. If it is possible, keep the lifting arm away from the puck when it is spinning and wait until you can slow it down to attack. Overhead hammer attacks can do some damage to a puck if you are able to aim well enough to hit the slow-moving center. Crushers and huggers should be prepared to take some

damage while grasping this spinner. The best bet for any robot against the hockey puck is to design so that there is at least one surface that can take the huge hits that will be dealt you by this opponent and only use your main weapon once the spinner has been slowed down.

Vertical disk and bar spinners will have a more difficult time with more opponents than will the hockey puck spinners. Still they do win matches in some spectacular ways. All the bots that face a vertical spinner should endeavor to stay away from the biting end. Attack the sides and rear. The driver of the vertical spinner should try to keep the spinner pointed toward his opponent. Because of the added difficulty in steering, maybe the best course of action will be to wait for the opponent to come to you, pivoting around a single point. Once the opponent is close enough, go for the attack.

The horizontal spinning disk and bar weapon is a bit more of an offensive weapon than the vertical spinner. After all, it is generally more maneuverable. You should always try to keep the spinner toward your opponent and go after their wheels and weapon.

The angled spinner has both the weaknesses of the vertical spinner and the benefits of the horizontal spinner. You should build this bot strong enough to bully your opponent in case your weapon gets disabled. Add a wedge or spikes, and make sure the drive train can handle pushing your opponent around.

A drum spinner is better than the standard vertical spinner because there is a wider point of attack, and it should be easy enough to build it so that it can drive while inverted. You are still vulnerable to lifters, hammers, crushers, and other types of bot, but you have a great chance to tear them up too. The spinning drum weapon is ideal for the builder who wants to build a strong bot with a decent chance of causing real damage.

Even though a big, massive saw blade is really cool looking, it is not likely to cause a lot of damage. You will still be vulnerable to every other type of bot. Use the same strategy as for

the vertical spinner. Wait for your opponent to come to you, and attack when it gets in range. That is, of course, if your opponent is not a hockey puck spinner. You want to charge and attack as soon as possible to keep the puck from reaching top speed.

Summary

This chapter covered just about every type of spinner in existence. If I missed one, there is an excellent chance that you will be able to apply parts of the strategies of the spinners I did mention. Obviously, the most effective and protected type of spinner is the hockey puck spinner. However, there is no such thing as a guaranteed winner. All spinners have to be able to withstand the amounts of force they inflict on their opponents. If they can't, they are dead in the arena. Even the hockey puck spinner driver must spend hours learning how to drive effectively.

12

General Design

Now that the theory part of the book is done, I feel this book wouldn't be complete without actual building examples to show how some of the weapons are implemented. In order to do that, you need to have an understanding of the major concepts in building. If you haven't read my other book, *Combat Robots Complete*, then you should. I go into a lot greater detail there on most of the following subjects. I'll also cover some material not found in the first book.

Seven Questions

When you start out designing your bot and weapon system you have to ask yourself seven important questions. Answer each one and you will have a much better chance at successfully completing your bot than you would if you simply start throwing parts together.

What Are the Rules?

You must know the rules for the competition that you wish to attend. Every serious bot competition has a Web site. Every Web site has a set of rules for the competition. If you plan to compete in some of the largest competitions, you might be in

luck. The Robot Fighting League (RFL) is attempting to band together as many competitions as possible. This includes putting them under a common set of rules. So, each competition should have similar rules. The main differences will be in the weight classes allowed and the amount of power allowed in spinner weapons. Some events may not allow spinner weapons at all. It mainly depends on the arena. So, if you build for an RFL event, the likelihood of being able to compete in several competitions with the same bot is very high. If you have a local competition and only plan to compete there, you still have to know their rules to build a legal bot.

What Materials Can I Use?

The materials in use today on combat bots are fairly easy to find in several places. New materials can be found in hardware stores, specialty stores, and on the Internet. Used materials can be found in junkyards, surplus yards, and on the Internet. The most common materials in use are aluminum, steel, titanium, and polycarbonate. Polycarbonate is better known as bullet-resistant glass. I say bullet resistant because it usually is not considered bullet proof until it gets a little thicker than is usually seen in a bot. Other materials can be found in bots too. Ultra-high molecular weight (UHMW) plastic is becoming more popular because of its strength and light weight.

There are rules that disallow certain materials. Those were covered in Chapter 2. The main idea is that you can't use anything that gives off radiation, toxic gases, or corrosive substances. Just make sure you know what is allowed before you try to use something. The last thing you need to consider about materials is whether or not you have the capability to put it to use. Nearly anyone with the right saw, a file, and some patience can turn any hunk of aluminum, steel, and polycarbonate into a robot part. The same can be said for any hunk of titanium, but the right saw and file are a lot less easily found. If you don't have access to a machine shop that can handle this material, you are better off spending less money on less-exotic materials.

What Strategy Will I Use?

Hopefully reading Chapters 4 to 11 will aid you in deciding on your strategy. I say that because strategy is tightly linked with the type of weapon you have on your bot. Each type of weapon has a different offensive strategy. Your opponent's weapon type will affect your strategy too. Even the type of competition will help dictate the strategy you follow. There are those competitions that are all fighting. There are also competitions that require you to navigate obstacles and traps. Some involve fighting and obstacles. Some new competitions are completely different from what you may be used to now. Be prepared to adjust whatever strategy you have to the environment your bot will live in.

What Weapons Will the Bot Have?

Combat robot weapons are very dangerous things. They are built that way because they have to inflict damage to what can pretty much be considered a tank. You should know that the forces required to destroy a tank are much greater than those required to do harm to a person. It's the same when you talk about a bot weapon. The slightest oversight can be your last. Do not try to build something that you have no knowledge about. Get some help. Start simple and work your way up.

I've said it before: No matter what weapon you put on your bot, there will be opponents that lose against it and opponents that win against it. Some of the most successful designs include two or more weapon types. The most common combination has some type of active weapon and a simple wedge. Almost every type of active weapon can be mounted alongside a wedge. In fact, the only one that can't is the hockey puck-style spinner. If there is any spot on your bot that isn't protected by an active weapon, it should be possible to add a wedge.

How Will I Control It?

There are several types of remote control systems used in today's combat robots. The most important thing to remember

when buying one is to make sure the system has enough control channels to control all the functions of your bot. If you're new to the sport, you should be keeping things simple. An on-off switch will activate just about every type of weapon. One control channel per drive side will also do the trick. This means a three-channel radio system should be sufficient for most bots. That's good news in a couple of ways. Three-channel radios are fairly cheap. Also, they come in a variety of configurations. If you have raced remote-control cars, you will be used to the pistol grip and steering wheel design. If you have flown remote-control airplanes or helicopters, you will be used to the standard airplane remote. For those of you who mainly play computer games, the IFI 900 Mhz system uses a conventional PC joystick. That brings us to the second consideration for you when buying a radio. Get one that you are comfortable with.

The thing to consider when deciding how to control your bot will involve the weapon type you choose. Spinner weapons are easy to operate with a single on-off switch. Most of the other weapon types can be operated the same way. However, the lifters, flippers, spikes, hammers, and whatnot can be very difficult to aim, especially while trying to position your bot to attack. Some radio systems can be connected to a trainer box so that a second person can operate the weapon, freeing the driver to concentrate on driving. Still other builders use two different radio systems to accomplish the same task.

Can I Pay for It?

I've been saying that you should start out small and simple with your first bot. The task of building a larger bot is daunting enough, but there is at least one other factor that makes my words true. Big bots are more expensive. It is true that you can dump a few thousand dollars into a 30-pound bot and only a few hundred dollars into a 200-pound bot, but the larger bot won't be too competitive. Even if you spend only a few hundred dollars on a large bot, other expenses will make my statement true. Flying or shipping a large bot out to a competition

will get very expensive. If you did only spend a few hundred bucks, you are going to need lots of spare parts that will cost more money. Overall, you have to be prepared to spend money on the bot, shipping, travel, hotels, and whatever else you can think of. This sport is expensive, and you have to be prepared for that fact.

Do I Have Time?

Designing and building combat robots is a time-consuming prospect. You will spend seemingly countless hours in the workshop preparing materials, bolting or welding them together, and wiring up all the electronics. That will be the bulk of your time, yet it will be spread out over late nights and long weekends. Another time taker is the event itself. Television events are notorious for taking between a week and ten days to complete. Even though television shows are dwindling, live events are coming into their own. At their start, live events usually lasted one or two days. Since they have become so popular, they are starting to take anywhere up to five days. My event is bursting at its seams when we try to cram it all into one or two days.

The Most Common Failures

There are lots of failures that happen to combat bots in the arena. They fall into four major categories. Speed controllers, motors, gearing, batteries, and other parts that are pushed beyond their limits fall into the first category. There are plenty of instances where motors are run on higher voltages than the factory specifies. This works to a certain point, but there are limits. You need to pay special attention to electrical limits. Running a 24-volt speed controller at 36 volts will always cause it to burn out. Running 300 amps through a 150-amp speed controller will do the same thing.

The next category is composed of the parts that aren't mounted correctly. Bolts can break easily if they are too loose or too tight. Wire connectors can come off wires if they aren't

crimped correctly or if the wire is put in a position to work loose. Chains can jump off sprockets, and gears can become unmeshed if motors, bearings, and supports aren't correctly mounted. Every bolt should be checked between every match to make sure everything is in working order.

"Unreliable parts" is another category of common failure. The most common of these is the home-built electronic board. I've seen lots of electrical engineers build their own speed controllers or radio interface boards. Invariably, these boards fail time and time again. I don't know how many times I've seen an engineer hooking up a meter or scope, trying to figure out what's wrong with his board while at an event. Eventually, if engineers work hard enough and spend enough time and money upgrading their boards, the equipment becomes reliable. The only problem with that is that there are plenty of boards available that have been tested and made reliable and cheap enough not to spend all that time and money on your own endeavors.

The last common failure category is bad design. TLC's Robotica show required bots to be able to cross swinging bridges, speed bumps, spikes, rollers, and sometimes even water or sand. Ground clearance was a major design factor for that competition. BattleBots and Robot Wars competitions are primarily made up of fighting on flat surfaces. Zero ground clearance is a major design factor for those events. Some events are combining fighting with maneuverability. Whatever the case, your design should be able to handle the task and achieve the goals set forth by the event organizer.

Motors and Drive Train

You aren't likely to enter a competition where the bots are driven by jet engines. The most common power source for bots is the permanent-magnet direct-current (PMDC) electric motor. Some people use internal combustion engines as power sources, but we'll keep it simple for now. The PMDC is called that because the magnets involved are permanently

mounted to the body of the motor. The armature of the motor has coils of wire that form an electromagnet.

When the electromagnet is energized, the magnetic fields of the coil and the permanent magnets are opposing each other. This causes the armature to move. If you place a series of electromagnets on the armature and turn them on in the right sequence, the armature spins. The commutator makes sure the coils are turned on and off in the right sequence. Brushes bring the current from the battery to the commutator. Changing the direction of the current flow through the electromagnets changes the polarization of the magnetic field and changes the direction of spin. **Figure 12.1** shows the basic PMDC motor.

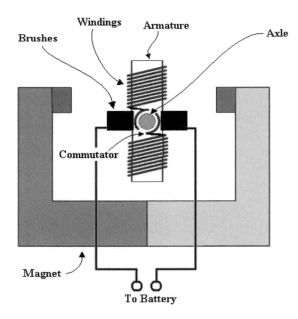

Figure 12.1

Basic configuration of the PMDC motor.

Current Draw

Given the fact that you have to apply a voltage to the coils of a PMDC motor, you should understand that current is also a factor. The amount of current depends on the amount and size of wire used to build the coil. The current drawn by a motor

is highest when you keep the armature from turning while applying voltage. This is called the *stall current*. The stall current is one of the most important factors to consider when picking a motor to use in your bot. Stall current affects the size of speed controller you can use, as well as the battery and wire size you need.

Most motors in use today are well documented by builders on their Web sites or by motor manufacturers. Sometimes the stall current and stall torque are printed on a name plate on the motor itself. If you can't find the specifications on the motor or on the Internet, you can still measure current draw and torque. Probably the easiest way to find the current draw is to use a D-cell battery, as shown in **Figure 12.2**.

Set your voltmeter to measure amps. Connect it, the motor, and the D-cell battery in series, as if you were going to power the motor with the battery. The motor, if large, won't move but the current from the battery will still flow through the coils. Record the current shown on the meter. Use Ohm's law to figure out the resistance of the motor's coils. Resistance (R) = voltage (V) / current (I). So, divide the 1.5 volts from the D-cell battery by the current measured and you will have the resistance you seek. Once you have the resistance, plug that value into another form of Ohm's law to find the stall current at the

Figure 12.2

Finding current draw with a D-cell battery.

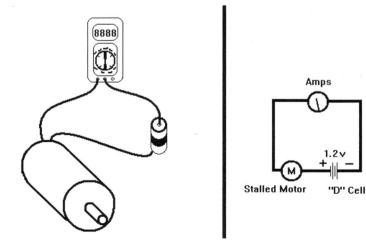

operating voltage. Stall current (I) = operating voltage (V) / coil resistance (R). Now you know the stall current. This will tell you what size of speed controller to use.

Torque

Stall torque tells you how strong the motor is. You don't want to design so that stall torque is required to move your bot. After all, stall torque is the amount of torque generated when the motor is powered up and the shaft is held in place. To measure stall torque, you need a vise, a scale, and a lever arm. Attach the lever arm to the shaft of the motor. Place the motor into the vise and secure it as shown in **Figure 12.3**. Apply full power to the motor and record the scale's measurement. Multiply the scale's measurement times the length of the lever arm and you have the stall torque.

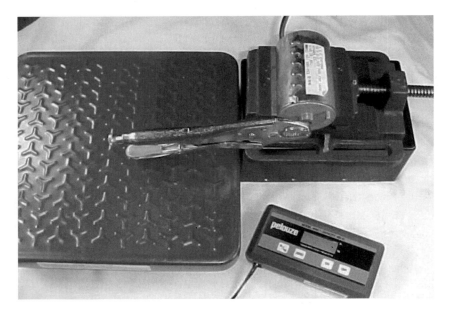

Figure 12.3

Finding torque with a scale.

Spinning Wheels

You can use stall torque and stall current to size your speed controller in a slightly different way than I've shown you. This method lets you get away with a slightly less expensive

speed controller because of the nature of a wheeled bot. If you use the right motor, the wheels of the bot will spin before they get a chance to stall. To figure this out, you have to know the torque constant. The torque constant (Kt) = the stall torque / stall current and is expressed in ounce-inches per amp. Be sure to specify the torque in ounce-inches. Now you have to know how much of the bot's weight is supported by a single wheel. A couple of caster wheels normally help support two-wheel-drive bots. Sometimes, as in the case of some thwack bots, the weight is supported by the drive wheels and the spinning tail. In any case, you should divide the total weight of the bot by the number of contact points on the ground. The result is the amount of torque it will take to spin a wheel. Multiply that by the wheel radius. Divide the answer by the gear ratio and you will have the amount of torque necessary from a motor to spin the wheel. If you divide that amount of torque by the torque constant, you are left with the amps that will be drawn by the motor when the wheel spins. Using the calculated amount of amps, you can now decide which speed controller to use. Be careful though. Any real damage to your drive train could cause a stall condition that might burn up your speed controller.

RPM

You should also know the top speed of your motor. This is measured in revolutions per minute (RPM). The easiest way to determine this is to look on the nameplate or the manufacturer's Web site. However, if you can't find either of these, you can measure it easily enough. If you plan to dig up motors for which there are no published specs, my recommendation is to buy an optical tachometer or tach (pronounced "tack"). People who fly model airplanes generally use these tachs to measure the RPM of their propellers.

These tachs rely on the propeller to block a light source. They count the number of times the light source is blocked out and convert that to RPM. You will have to fasten a mock propeller to the motor shaft and give the tach a light source. Do not use

an AC-powered light source. Believe it or not, AC-powered lights turn on and off sixty times a second. This is fast enough for your eye not to detect it, but not so fast that the tach can't. A flashlight is probably your best bet.

Knowing the speed of the motor will tell you the speed of the wheels. If you are using wheelchair motors, there are no other calculations to make. If you are using a standard motor and a separate gear or chain reduction, you will have to divide the motor RPM by the reduction ratio to find the wheel RPM. Use the current draw of the motor to choose your speed controller. Use the wheel RPM to see if your motor will move your bot fast enough. If the RPM is too low, you need to find a new motor and recalculate for the current draw.

Chains and Sprockets

You might choose to use wheelchair motors exclusively in your bot designs. This is perfectly fine. However, I like the challenge of building a homemade gear system. I don't suggest it for the new builder. An easier way to get started is with sprockets and chains. **Figure 12.4** shows the ups and downs of positioning a chain drive system. Sprockets and chains are

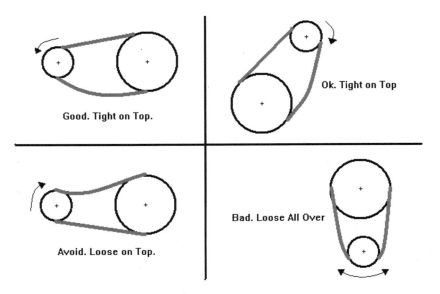

Good. Tight on Top.

Ok. Tight on Top

Avoid. Loose on Top.

Bad. Loose All Over

Figure 12.4

Chain positioning.

fairly forgiving. They can sustain a little bit of misalignment. Notice in the figure that the chain always has a loose side and a tight side.

The loose side is always the side that the drive sprocket is pulling toward. You have to remember that, in a bot, the loose side is always switching because the motor is always changing spin directions. You can take the slack out of the chain with an idler sprocket, as shown in **Figure 12.5**. An idler sprocket is just a sprocket that spins along with the chain. Its position should be adjustable so that, as time goes by, any slack can be removed. **Figure 12.5** shows two methods of using idler sprockets. The method on the left is using a single idler sprocket. This can be adjusted to keep everything tight or can be spring mounted. The right side of the figure shows how to use double idler sprockets. Using this method will help ensure that the chain stays on as long as the sprockets are aligned correctly.

The best way to get your sprockets in line with each other is to place a straightedge across both of them as shown in **Figure 12.6**. Make sure there is no light shining between the sprockets and the straightedge. This will be easier to see if the straightedge sticks out past both sprockets by a few inches.

Figure 12.5

Idler pulleys.

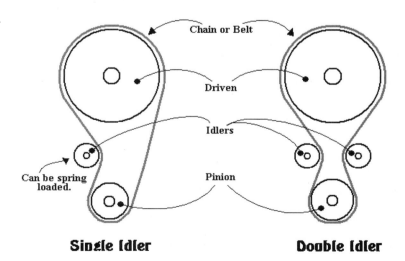

Misaligned
Gap between
sprocket and
straight edge.

Aligned
No gap between
sprocket and
straight edge.

Figure 12.6

Checking
sprocket
alignment.

Figure 12.7 shows you the basic belt and chain length formula. Use this to figure out how long your chain should be, and you'll have less slack to worry about. If you are simply transferring power to a set of front wheels by using a one-to-one reduction ratio and two sprockets of the same size, then you can use a simpler formula to calculate the length. Length = twice the center distance × the circumference of one sprocket. Circumference, if you didn't know equals π times diameter.

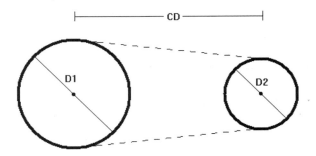

Figure 12.7

Basic belt and
chain length
formula.

$$L = 2CD + \frac{3.14 * D1}{2} + \frac{3.14 * D2}{2}$$

Wheels and Walkers

Your wheels are your next concern. Many builders prefer pneumatic wheels. Fat, knobby tires give good traction and come in lots of sizes. Cautious builders take an extra step and either purchase foam-filled tires or have the regular ones filled later. This will help your bot survive in arenas with big saw blades in the floor. It will also help against spinners or any other type of bot that can flatten regular air-filled tires. Many other builders prefer solid wheels. Probably the most-used solid wheel is the Colson caster wheel. This has a solid plastic core with a pliable rubber tread. These wheels can take a beating without going flat. However, it's only a matter of time before they get destroyed if you leave them unprotected against your opponent.

Wheels are important in getting your bot to move, unless you choose to build a walker or a shuffler bot. Everyone knows what a walker bot does. A shuffler bot may not be as self-explanatory. Some competitions give a weight bonus to walker bots because of the added complexity and diminished structural integrity. The bonus is meant to be an incentive to build a cool walking machine. However, some builders thought up the shuffle bot. It doesn't roll like a rolling bot but it doesn't walk like a walker either. Sets of long feet are rotated with a cam mechanism so that one is placed in front of the other. It is possible to build this shuffle mechanism to be every bit as strong and reliable as a wheeled drive train without the weight bonus. Because of this, the added weight can be used to beef up the armor and weapons of the shuffle bot. This is exactly the reason why new rules have been drawn up to differentiate between shufflers and walkers.

Modeling

Once you figure out the main parts of your bot and the overall type of bot you want to build, you need to make sure you can stuff all your components into the spaces allowed by your design. I have, and you can, started building without any kind

of plan other than the glimpses of light through the fog in my head I like to call creativity. If you are talented in that aspect, go for it. However, most of us need a plan so that we don't have to change fifty things before we even get the wheels mounted.

If you are used to working with computers, a simple computer aided design (CAD) program can do wonders for your planning abilities. You draw all the parts of your bot to scale and place them in three-dimensional space on the computer screen. That way you know where everything goes and whether or not there will be a problem when it comes to physically mounting them. There are several cheap CAD systems and several expensive CAD systems. I like Rhinoceros (Rhino3D). Rhino3D is a mid-range CAD system when it comes to price. You can get it for less than $1,000. Some of the high-end software is twenty times that. However, for the price, you get everything you will need to design some very complex machinery. It isn't too difficult to learn either.

If you don't want to spend time learning software and are more of a hands-on type of person, physical modeling might be your thing. Erector sets can aid you in building strong frames. Lego sets can really make it easy to test new bot designs. Even laying all the parts on the floor will help you figure out whether or not everything will fit the way you think it will. The point is, a little planning will go a long way when you start to build something. It will also save you some money.

Frames

There are a couple of different types of frame that I'd like to talk about. The unibody and the stick frames are probably the most common in combat robots. The unibody is essentially one piece of material formed to hold the motors, batteries, and other parts of the bot. The stick frame is normally made up of pieces of either round or square tubing.

The Unibody

Some unibody frames start out as a huge chunk of material. This can be anything from steel to titanium though aluminum is probably the easiest to work with. The builder uses a milling machine to "hog" out most of the material, leaving the shell, motor mounts, and whatever else the design calls for. After everything is mounted, a lid is screwed down to the top to seal everything up. **Figure 12.8** shows a CAD drawing of a unibody frame that will be made in this way.

Figure 12.8

CAD design of a unibody frame.

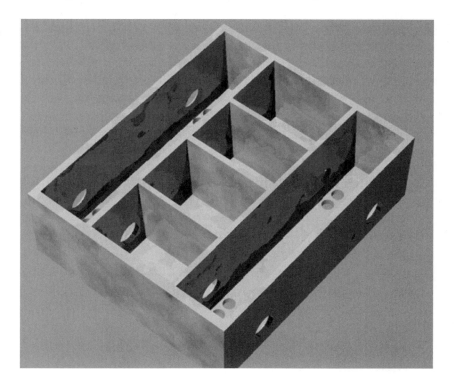

Figure 12.9 shows another type of unibody frame. This one is made of aluminum plates that have had slots cut so that they fit together, much like one of those three dimensional puzzles. After every slot has been cut and every hole has been drilled, the plates are fitted together. Sometimes the builder will bolt the frame together before welding, it but once welded, it is completely solid and very sturdy.

Figure 12.9

Unibody frame of Phrizbee. (Courtesy of Brian Nave and Team Logicom www.team logicom.com)

Stick Frames

Figure 12.10 shows the stick frame from one of my bots, called the Six Million Dollar Mouse. Stick frames are very strong. They're also the easiest to build for a beginner without many tools. You can build a stick frame with a saw and a welder. The most common type of material used to build a stick frame is square tubing (steel in the heavier bots and aluminum in the lighter ones). Square tubing is probably the easiest to work with since body panels, motor mounts and other components can easily be mounted to its flat surfaces.

Round tubing will give you a stronger frame at the expense of some ease of assembly. You can see round tubing put to use as bicycle frames, race car frames, roll cages, and sports equipment. It is not impossible to mount things to round tubing but it is a little more time consuming. It is even more time consuming to weld the tubing pieces together. **Figure 12.11** shows two pieces of round tubing that will be welded together. Notice the round cut on the end of the short piece. This cut is made with a coping saw. Once the main cut is done, you can

Figure 12.10

Stick frame of
Six Million Dollar
Mouse.

use a file to smooth it out and fit it perfectly against the second piece of round tubing. If you didn't make the cut, the pieces would only be held together by a small amount of welding and would surely fail upon testing.

Frames in General

When you build a bot frame, you want it to be as strong and light as possible. The one-chunk, unibody frame works well for

Figure 12.11

Cutting a round
tube for a flush
weld.

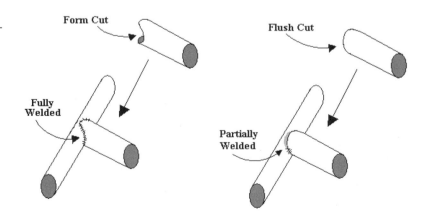

small bots, but I wouldn't want to build anything heavier than 30 pounds that way. Even at 30 pounds, the starting material for one of my 30-pound bots would weigh about 95 pounds. That is at 3 inches thick. I couldn't find anything over 1-1/2 inches thick and an 18-inch square piece cost over $600. So you see, I'd have to spend over $1,200 just to get material that wouldn't work unless I made a top and bottom half of the unibody. Even then I'd be turning 90 percent of the material into chips on the shop floor. I would come out a lot cheaper with a lot less work by making the frame out of plates and welding them together with some well-placed supports. It is true that I might have been able to find some suitable material for a lot less in a well-stocked scrap yard. The scrap yard is a great place to find materials. It's also a great place to see how engineers build strong structures. You might not believe the amount of incredible "stuff" that can be found at a scrap yard. I'm not talking about the junkyard down the street with all the cars piled up on top of each other. I'm talking about the scrap yards that buy junked industrial equipment from large companies.

Figure 12.12 shows some of those well-placed support brackets I just mentioned. Every frame can use some type of strengthening support. If you didn't get any ideas from the

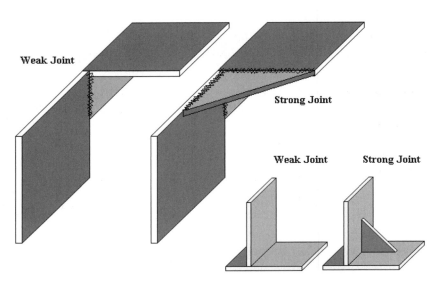

Figure 12.12

Frame support brackets.

scrap yard, or you don't have one close enough to visit, then you can always pay attention to several letters of the alphabet. The letters I, L, and T can easily be found if you pay attention to even the simplest of available materials. Steel I-beams are very common. Aluminum angles, or Ls, are easily found. Triangles are probably the most structurally sound and can be found in the letter A. In fact, the most structurally sound frame you can make is one where you join the sides of four equilateral triangles. That's probably why the Egyptian pyramids are still hanging around.

The combination of strong and light is sometimes elusive when it comes to bot frames. It might seem illogical, but hollow tubing is stronger than solid bars. It is also lighter, but that should be evident. When you start designing your bot you should have your motors, batteries, gearing, wheels, electronics, and whatever else decided upon. You should know their combined weights. This should give you an estimate of how heavy your frame and body can be. Imagine that your frame can weigh about 20 pounds. Also imagine that you plan to use 1-inch square tubing with 1/16-inch thick walls. This is fairly light and strong material for a robot frame. In fact, it is what I used to build the frame shown in Figure 12.10. This material weighs about 2 pounds per foot. Simple math shows that you can use about 10 feet of this tubing and be within your expected frame weight. Now you can either model the frame in a CAD system or make cardboard cutouts of 1-inch tubing to make sure it can be done within 10 feet of material. If you find that the frame you want cannot be made in that amount of material, you will need to rethink your desired frame or frame material.

Specific Structures

I want to talk a little about individual bot structures. Mainly I want to include the spinner support, hammer support, and wedge support. Within these three types of structure, you will find plenty of information to build any other type of bot that you like.

Spinner structure. Spinner frames need to take a lot of beating. They inflict massive amounts of damage on their opponents but the spinner experiences the same forces. Large bearings are the ticket to supporting the spinning mass whether it is spinning horizontally, vertically, or at a diagonal. I don't recommend cast iron housings for the bearings. Cast iron is porous and brittle. You can use a hammer to smash cast iron bearing housings. It should be obvious that a bot can do the job more easily. Aluminum bearing housings will work fine for the lighter bots. Some people are even using a very strong version of plastic. Steel housings, as shown in **Figure 12.13**, are recommended for the heavier spinner bots.

Figure 12.13

Large bearing with steel housing.

The weapon of any spinner is its spinning mass. That mass must be sufficiently structurally sound for it to be effective against your opponents. Spinning disks and bars are probably the easiest to make. You just have to pay attention to how the disk or bar is mounted. Simply sliding it onto a shaft and bolting it to a hub doesn't really get the job done. Bolts aren't made to take the shear forces that will be generated. A large key, as shown in **Figure 12.14**, will help withstand the forces and keep the spinning mass from spinning around the shaft. If you can't get a key into the shaft or spinning mass, at least use

Figure 12.14

Key in a spinning
weapon.

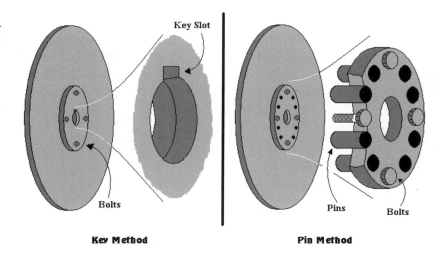

Key Slot

Bolts

Pins

Bolts

Key Method   **Pin Method**

hardened pins, also shown in **Figure 12.14**. If you go with the
pin/bolt combination you will need to make sure the disk cannot travel up and down the shaft. Shaft collars can do the job.
These are simple rings of metal that have some method of
clamping down on the shaft. Some collars just have a setscrew
that tightens down on the shaft itself. These are fairly unde-
sirable since they only hold the shaft in one spot and can
cause marring in that spot. Other shaft collars are split on one
side, with a bolt running through the split. When the bolt is
tightened, the entire collar's inside diameter shrinks and
holds onto the shaft without marring.

The hockey puck spinner shell is probably the most difficult
spinning mass to build. Depending on the size of your bot,
you might be able to find something off the shelf to modify
and form your shell. Some builders have used cooking woks,
heavy saucepans, and even pipes. But, if you plan to build a
large bot, you probably won't find anything lying around. In
this case you will need to find a machine shop that can bend
plate steel into a circle. For the big bots, you should use 1/4-
inch-thick material. Even then you might be hammering out
some dents in between matches.

The drum spinners can be constructed in the same way,
though it might be a little easier to find something that already

resembles your planned drum. Though I thought the material was a little thin, I've seen drums made from small steel barrels. Some of the better ones can be made from old propane or natural gas tanks. Just remember that the thicker the material is on the outside wall of the drum, the better.

Wedge structure.   Probably the most important part of building a wedge bot is the face and point of the wedge itself. Some bots use a simple flat piece of material that is hinged at the mounting point. **Figure 12.15** shows this type of wedge.

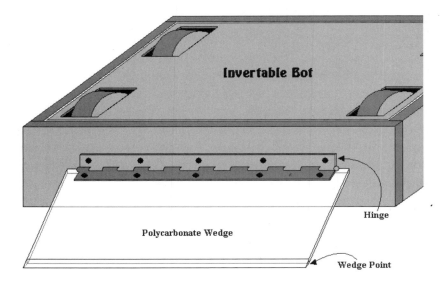

Figure 12.15

Flat wedge.

The point of this wedge, the part that meets the ground, should be as thin as possible in order to get under the wedge of an opponent. Thin aluminum or steel can be deformed quite easily when slamming into an opponent, especially if your bot isn't lined up properly to get under its opponent. Thicker material can be used as long as you try to sharpen the point. You can also try to use some stronger, thin material such as titanium. You'll have an easier time if you mount the wedge face at some angle less than forty-five degrees. In fact, the lower the wedge, the easier it will be able to get under your opponent. The flip side to that is that a wedge that lies

flat on the floor doesn't exactly do anything to get your opponent off its wheels.

On bots that are able to run when upside down, the flat wedge will be ineffective if not mounted correctly. You should mount your wedge on hinges about halfway up the sidewall of your invertible bot so that when it is flipped, the wedge can flop down and remain effective. Of course, this position might be too low to do any good if your bot has a very low profile.

Figure 12.16 shows the skirt wedge of Dr. Inferno Jr. This wedge gets its name from the fact that it skirts the perimeter of the entire bot. Some bots skip the front skirt section and add an active weapon. Small bots normally use the skirt wedge because it makes the overall bot footprint much larger, but there's no reason, other than arena door size, that you couldn't use it on a big bot. The skirt wedge is usually a variation on the flat wedge. Not a rectangle shape, a single part of the skirt has two parallel sides and is angled so that it meets two other parts of the skirt to give full coverage around the bot perimeter. Notice that Jason has hinged the bottom inch of this skirt. This helps the bot get over bumps as well as making it more difficult for an opponent to get under.

Both the skirt and the flat wedge should be mounted on hinges for maximum utilization. The skirt wedge has normal-

Figure 12.16

Dr. Inferno Jr's skirt wedge. (Photo courtesy of Jason Bardis www.infernolabs. com)

ly been used on bots like Biohazard and Dr. Inferno Jr. I'm not saying that a skirt wedge won't work on an invertible bot, just that I've seen it mainly in conjunction with active weapons on bots that can't run while upside down.

Figure 12.17 shows what I call a compound wedge. Unlike a flat wedge, the compound wedge starts out at a low angle and slopes up steeply. I like this wedge better on large robots rather than small ones. On a small robot, there isn't a lot of room for the low angle to develop into a steep one. On larger bots, the wedge has time to stay low long enough to get significantly under the opponent. Once it is under the opponent, the steep angle helps pick the opponent up higher, eventually tipping it over completely. Wedges like this can help deflect spinner attacks too. I don't have a magical formula that draws up a perfect compound wedge but I'm sure they are out there. All you have to do is look around. I've seen some great compound wedges on bulldozers, snowplows, and other industrial instruments of pushing.

Hammer structure. Just like spinner bots, hammer bots have to be able to absorb the same forces they use against their opponents. There are a few ways to dissipate this energy without tearing apart your bot.

Figure 12.17

Compound wedge on MiniRip.

Figure 12.18

Simple
mechanical
energy
absorption.

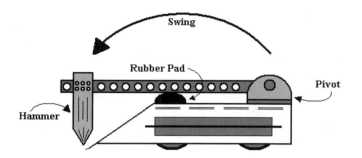

Using large rubber pads, strong belt restraints, or a combination of the two can attain simple mechanical energy absorption in a hammer weapon. In **Figure 12.18**, the hammer simply slams into the rubber pad at the end of its swing stroke. This only happens when you miss your opponent. Some event organizers might even require that your hammer/axe weapon not be able to hit the floor when you miss.

Figure 12.19 shows another method for dissipating energy. This time we've installed a heavy-duty belt that will extend to its total length before the hammer reaches the end of its stroke. The belt can be made of several different materials. Some people use kevlar. Some people use springs or rubber cords. Some even use standard seatbelt material. The main point is that you have to be sure to attach both ends very securely. Be careful how you mount this type of mechanism. You don't want to be ripping your bot apart from the inside.

Figure 12.19

Simple
mechanical
energy
absorption.

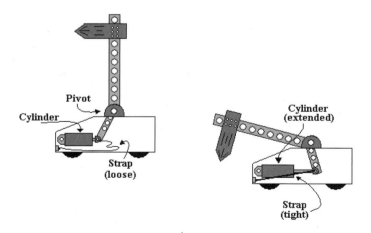

There are other simple ways to absorb energy during a miss, but my favorite one is really only available to the builders who chose to use a pneumatic system. Since air acts as a spring inside a confined space, you can use that spring action in the exhausted side of the pneumatic cylinder that moves the hammer. **Figure 12.20** shows a cutout of a pneumatic system using this method to slow the hammer swing when it gets close to the end of the stroke.

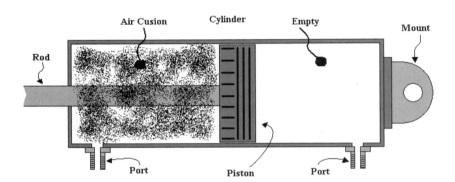

Figure 12.20

Pneumatic system exhaust buffer.

This system is a bit more complicated, since it requires timed electronic control over the input and output air supply lines to the cylinder. As gas is rushing into the input side, starting the movement of the hammer, gas is also being let out of the exhaust side. The gas on the exhaust side is usually vented to the atmosphere. This means that the pressure formed on the exhaust side is equalized with normal atmospheric pressure. This does not mean that there is no air left. The amount of buffer you want in your system determines when you need to close the exhaust valve. Since the exhaust gas will vent almost instantly, that amount of time has no real effect. If you really wanted to start slowing your hammer when it reached half its stroke, you would close the exhaust valve soon after the hammer reached that position. That really isn't likely. But if you want to figure out the best position to start the slowdown, you should start somewhere near that point and change the timing until you get the results you want.

I'm sure there is some complicated mechanical way of changing the valve timing so that you get this type of buffer action.

In fact, there are hose connectors that can vary the amount of airflow. You can use these in a slightly different way with success and, in truth, they aren't that complicated. However, I'm a tech freak, and I prefer software-based solutions. I believe they give more flexibility and can be made to be just as reliable but with more failsafes. Several teams make use of the IFI radio system to program the timing of the valves. Some teams use PIC microcontrollers. Whatever your brand of electronics, the programming shouldn't be too difficult to develop.

Summary

This was a big chapter but then again, general design is a big subject. First off, you have to answer the seven important questions of robot design. While answering those, you should keep the most common failure list in mind. Once you have a general bot design, you should try to model it in a couple of ways to make sure it will work and to make sure everything will fit. Building small working models will accomplish the first task. Drawing your bot to scale in a CAD system or constructing an actual-size model from cardboard will make sure everything will fit. Once you know these things, it's time to spend some effort designing the frame. There are several tips in this chapter that will help you with that. They range from general frame tips to specific ones dealing with spinners, wedges, and hammer weapons. The overall concept of this chapter is to get your toes wet with the design details.

The next chapter will focus on specific examples of radio control methods used in bots with active weapons. While most of the circuits are the same when it comes to simple on/off controllers, there are several different devices available that have different features that might be desirable to you. Also, there are current requirements that must be examined and dealt with. The electronics of the next chapter will be discussed on a basic level that should be easy to understand if you can think of a line as a wire and a box as a remote control interface, speed controller, or radio receiver. You'll see what I mean.

13

Electronics and RC Equipment

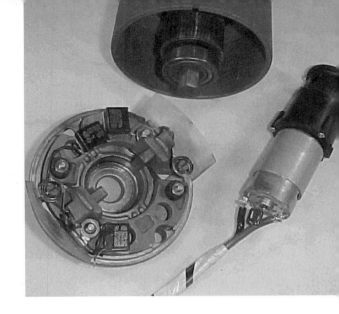

The last chapter talked about general mechanical design and construction of robot combat weapons. This chapter will go into some detail on the electronics used in these bots and weapon systems. For the most part, all the electronics can be watered down into simple electronic devices that can be bought "off the shelf" in online robot combat stores.

One of the pioneers in robot combat electronics is Dan Danknick of Team Delta (www.teamdelta.com). Along with being an exceptional bot builder, Dan is an embedded software engineer who also happens to run the oldest online store for robot parts. He started out by designing an interface for 4QD speed controllers and a radio receiver. Soon after that, he developed a simple electronic switch that could be controlled with the radio. Dan would study his and other builders' needs in order to come up with new products. Eventually he was offering a full line of electronic devices that made life a lot easier on the bot builders.

Team Delta is probably the most successful combat robot parts store in existence because of the excellent products and even better customer service. Dan routinely sponsors robot fighting events and in some cases he even flies out to lend a hand as an event staff member. You might think that these things are

why I'm including a lot of his products in this book but that is not the reason. The real reason is that I have used his products in every bot I have ever built and not a single one has given me a problem if I used it correctly. Even when I didn't use it correctly and couldn't figure out what the problem was, Dan quickly figured it out and I was on my way to an event.

Getting Started

Before we go into specifics on how to use several of these interfaces, you'll need to be familiar with some very simple schematic conventions. Schematics are made up of wires and symbols. Symbols represent the different components of a circuit. Wires connect those components to each other. **Table 13.1** shows the figure designator, component name and component description for all the component symbols listed in **Figure 13.1**. Study the descriptions and symbols so that you know what each one is and does.

Table 13.1

Schematic Figure Designations and Descriptions

Figure	Name	Description
A	Wire	Connects different components. Thicker wire can handle more current. Thick lines indicate that you should use heavy-gauge wire.
B	Connection	Wire connections are indicated in this book with a dot.
C	No Connect	A non-connection between two wires that cross is indicated without a dot.
D	Terminal	An input or output of a device.
E	Polarized Terminal	An input or output of a device that should be positive or negative. This is indicated by the plus or minus sign next to the terminal.

(continued on next page)

Table 13.1

Schematic Figure Designations and Descriptions (continued)

Figure	Name	Description
F	Supply Voltage	Indicates that this line is connected directly to a positive terminal of a battery.
G	Ground	Indicates that this line is connected directly to a negative or ground terminal of a battery.
H	Battery	Shows the positive and negative terminals of the battery. Will also show the voltage of the battery. Sometimes one battery with a higher voltage number will be used to indicate what is really one or more lower voltage batteries wired in series.
I	Switch	A device to make or break a circuit. Includes all manner of switches. Shown is a single-pole single throw switch (I-a) and a double-pole single-throw switch (I-b).
J	Coil	A winding of wire used to create a magnetic field. Coils, as referred to in this book, are used in relays and contactors or as solenoids to electromagnectically actuate the switch(es) inside the relay, contactor, or solenoid.
K	Contactor or Solenoid	An electromechanical switching device for controlling a high voltage or current with a low voltage or current. Usually has a single electromechanical switch and is normally rated to handle higher currents.
L	Relay	An electromechanical switching device used for controlling a high voltage or current with a low voltage or current. Can contain many electromechanically actuated switches. Not usually rated to handle higher currents, but can be used to actuate a contactor or solenoid.

(continued on next page)

Table 13.1

Schematic Figure Designations and Descriptions (continued)

Figure	Name	Description
M	MOSFET	A solid-state electrical switch used in speed controllers.
N	Capacitor	Used to limit motor noise. An electricity storage tank.
O	Diode	Used to limit relay, contactor, and solenoid generated spikes. Allows electricity to flow in one direction only.
P	Resistor	Used to limit current. Sometimes used in conjunction with a meter to calculate current (in that case it is called a shunt).
Q	Motor	Drive or weapon mechanisms. Could be used in place of the actuator icon. Reverses spin direction when the polarity of the applied voltage is switched.
R	Actuator	Drive or weapon mechanisms. Should not be used in place of a motor icon.
S	Meter	Used to measure many electrically related values, including voltage and amperage. The arrowhead leads indicate test connections.
T	ECL	Emergency cutoff loop. Kills power to the entire robot.
U	Antenna	Helps transmit and receive remote control signals.
V	Any Device	Usually a block with labeled terminals. Takes the place of anything that makes the schematic too complicated to draw. Also takes the place of standardized devices where it would not be prudent to include the exact wiring diagram or for which you do not have the exact schematic.

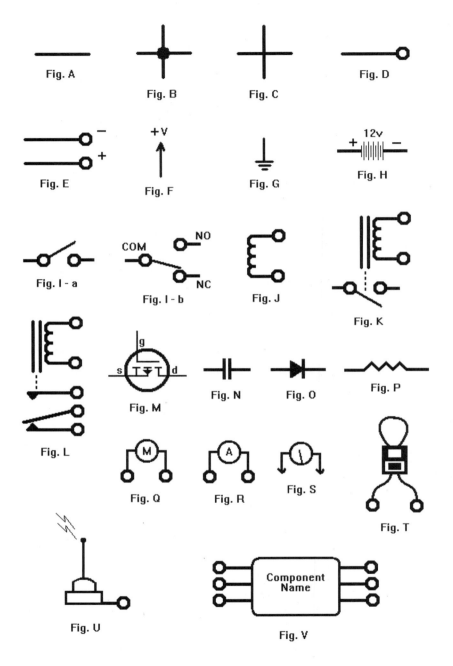

Fig. A

Fig. B

Fig. C

Fig. D

Fig. E

Fig. F

Fig. G

Fig. H

Fig. I - a

Fig. I - b

Fig. J

Fig. K

Fig. L

Fig. M

Fig. N

Fig. O

Fig. P

Fig. Q

Fig. R

Fig. S

Fig. T

Fig. U

Fig. V

Figure 13.1

Component symbols used in this book.

Figure 13.2

Basic diagram of a remotely controlled motor.

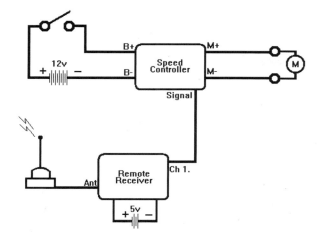

Figure 13.2 shows a motor that is remotely controlled with the standard radio and speed controller. The motor and speed controller are powered by a twelve-volt battery. Main power is controlled by a switch. The remote receiver is powered by a five-volt battery (actually it's a 4.8-volt NiCd pack) and uses a single channel to talk to the speed controller. This is probably the simplest setup you will see. It is also almost as complicated as it gets.

Wire Size

Wires carry current to components in the circuit. Current is basically electrons, but even something as small as electrons can cause heat if you try and pump enough of them through a wire. Just like a pipe, the larger it is, the more water you can pump through it. With wire, the larger it is, the more current you can pump through it. If you try to pump too much current through a small wire, the wire will heat up. It may heat up hot enough to burn the insulation and cause a short. A short is the condition that happens when two wires touch each other when they aren't supposed too. When this happens, you can pretty much count on melting your batteries. You want to use the correct size wire when building your bot. I have a simple rule of thumb when it comes to wire size. If I'm going to be

running more than 10 and less than 60 amps I use 12 gauge wire. If I'm going to be using more than 60 and less than 100 amps I use 10 gauge wire. If I'm using anything over 100 amps I go with 4 gauge wire. So far, I haven't had a wire burn but I also haven't drawn more than 300 amps or so. Anything over that and you might want to go with a bigger wire.

Startup and Shutdown

As a matter of safety, I need to mention something about turning off a combat robot. Remote control systems are truly made for toys. With the exception of the IFI radio system, all radio systems in use today were designed to get a toy car, truck, airplane, or helicopter around a circle. Even the IFI system was designed for robots that do not concentrate on destructive prowess, even with the radio failsafes programmed correctly, you can create a dangerous environment when turning your bot on and off. It is widely believed that the safest way to operate your bot is to follow a few simple steps in order. To turn your bot on, do this:

1. Turn on the transmitter.

2. Make sure all the trim settings are correct and joysticks or wheels are centered.

3. Make sure all the switches are set so that weapons are deactivated.

4. Turn on the receiver and wait several seconds for it to settle and get the signal.

5. Turn on the main power to the robot.

If you are at a legitimate event, they should have a frequency clip for each available frequency. Make sure you have your clip before you ever turn your transmitter on. Make sure you return it after you are done with your fight or done testing. Others might be on that same frequency and need the clip to go in the ring or do some testing of their own. If you have one, and I highly recommend it, you should use your emergency cutoff loop to cut the main power to your bot.

The emergency cutoff loop is in the rules of some but not all events. The main reason for it is so that you have a way to easily and quickly kill the power to the bot. It was named incorrectly. If the bot is uncontrollable in the arena and not putting anyone in danger, you should let the batteries run down before trying to grab the loop. Under these circumstances, you should never just run in and try to kill the bot. The opposite is more likely to happen. **Figure 13.3** shows a picture and a schematic of an emergency cutoff loop installed in a simple system. The only thing missing from the schematic is a speed controller. The loop itself is a loop of wire that is large enough to carry the total current load of the whole bot. You disconnect power simply by yanking the connectors apart.

Figure 13.3

Picture and schematic of emergency cutoff loop.

 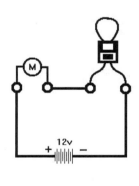

Relay Control

Most builders use a relay, an electromechanical switch, or a contactor, a big version of a relay, to control power to their weapon. Contactors are used to turn on large motors. Relays are used to turn on small motors or pneumatic solenoid valves. **Figure 13.4** shows the setup for a simple motor control. You see the motor, the 24-volt motor battery, the contactor, the contactor battery and the RC interface. We'll get to the interface in a little while. For now, concentrate on the diode

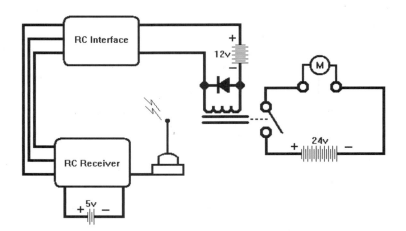

Figure 13.4

Standard contactor control.

that is connected across the leads of the contactor. Diodes allow current to flow in only one direction. In this example, the diode is connected so that it conducts current only when the relay is supposed to be off; otherwise it would create a short circuit and the contactor coil would never turn on. The reason for the diode is that the coil of the contactor generates an electrical spike when power is removed. The spike is in reverse polarity with respect to the electricity that originally powered the coil. Because of the reverse polarity, the diode conducts the spike. The spike is then dissipated back into the coil instead of back into the electronics of the RC interface. Diodes for use in this manner are usually black with a white band around one end showing the direction of current flow (from positive to negative, white side is negative). The straight line in the diode's schematic symbol represents the white band. You can find diodes at any number of places, including Radio Shack. Make sure they are rated for at least three times the operating voltage.

Radios

One of the most important parts of your bot is the remote control unit or radio. There are a few different kinds of radio to chose from: AM, FM, and digital. These only resemble the car radio in the way they encode the signal. You won't be picking

up your favorite rock station on them anytime soon. AM radio controllers are not suitable for combat robots. They are way too susceptible to interference. Every competition that I know of requires you to use an FM or digital radio. Even the small 1-pound bot competitions seem to be leaning into requiring a "real" radio.

FM controllers are of three styles: PPM, IPD, and PCM. PPM is a standard, run-of-the-mill, radio controller. With this type of radio you have to use separate failsafe mechanisms. The IPD is pretty much the same as a PPM radio only with minimal failsafe programming. Most events will place guidelines on which weight classes can use these two types of radio. Usually they are only allowed on the smaller bots. The PCM radio is computer controlled. Actually the IPD radio is too, just not as extensively. The PCM radio codes each packet of information sent to the receiver. This makes it easier for the receiver to distinguish between packets sent on its frequency and packets sent on close frequencies. The PCM is by no means completely reliable but is the best solution for the price. You can pick up a Hitech brand PCM radio system for a little over $300. **Figure 13.5** shows a standard joystick radio along with a wheel-type radio.

Probably the most reliable, yet expensive, radio system is the IFI robot controller. Made by Innovation First Inc.

Figure 13.5

Joystick and wheel transmitters.

(www.ifirobotics.com), this radio operates on the 900-Mhz frequency just like a lot of cordless phones. The IFI is basically a small computer with a radio modem attached. The receiver is also a small computer with a radio modem attached. The computer runs a program that reads switches and joystick positions, encodes them, and transmits them. The receiver decodes the information and sends the commands to the computer. The computer controls the speed controllers and whatever else is attached. The special features of the IFI system are how much you can control with very simple programming knowledge and the fact that the receiver can send information about the bot back to the driver. **Figure 13.6** shows an IFI robot controller.

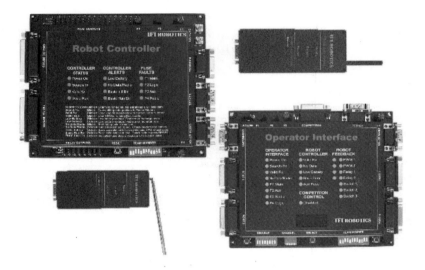

Figure 13.6

IFI robot controller.

Either way you go with radios, you're going to have to pass safety inspection. This means you need to read up on how to program those failsafes and make sure they are working before you get into the arena. Every radio has an operator's manual. You'll get one if you buy new equipment. You might not get one if you buy used equipment. If you don't, you can usually get in touch with the radio's manufacturer to get a copy. If that fails, try asking some builders if they use the same radio and can get the manual copied. If you still have no luck, you probably have a really old radio and should just try to find a newer one.

The biggest problem you will have with your bots will be radio reception. The three main causes of problems are RC battery power, a badly installed antenna, and radio frequency interference. You can take some small steps to ensure these problems don't bite you when it comes time to face your opponent.

The four-cell, NiCad, receiver battery pack is not made to source a whole lot of current. It is made to power the receiver and a few small servos. Nowadays, the bot builder is hanging large relays off the same battery supply. This will certainly drain the battery pack a lot quicker than a few servos. As long as the pack will supply the current for the duration of the match, there are two remedies to this problem. The first is to make sure you charge the pack after each fight. You might want to keep a watch on the transmitter batteries too. They won't die as quickly as your receiver pack, but they do run down and will keep your bot from working correctly when they are too low.

The second way to solve the battery power problem is to eliminate the receiver pack all together. The battery eliminator circuit is a DC-to-DC voltage converter as shown in **Figure 13.7**. You hook it up to the receiver just like a battery. Then you hook it up to your main drive power batteries. The converter changes the higher voltage to a lower voltage suitable for running the receiver. Now, your radio will stay on as long as your

Figure 13.7

BEC. (Courtesy of www. TeamDelta.com)

bot has power left in its main batteries. I'll go into more detail on this a little later.

The next radio problem you will experience has to do with the antenna. On stock receivers, the antenna is a long wire that just hangs off the back end. Depending on the frequency of your radio, the length of the antenna has to be somewhat to specs. 72-MHz radios need an antenna that is about 41 inches long. A 75-MHz antenna should be about 39 inches long. Now, running about 40 inches of wire somewhere in your bot could be a little troublesome. I prefer to change the long wire to a loaded whip antenna, as shown in **Figure 13.8**.

The third problem that you will likely face with radios has to do with self-generated interference. When motors spin, their brushes make and break contact with the commutator. This causes sparks and sparks cause RFI or radio frequency interference. There are a few things you can use to fight RFI problems.

The first and foremost weapon for killing RFI is the capacitor. Capacitors either block or allow electricity to flow through, depending on the frequency of the electricity and the size of

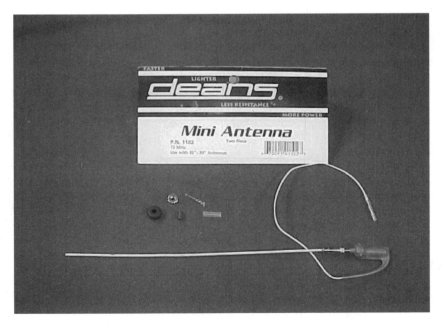

Figure 13.8

Dean's base loaded whip antenna.

the capacitor. The smaller the capacitor, the higher the frequency it will swallow or make disappear. It takes a good bit of expensive and specialized equipment to detect the frequency of the noise generated by motors, because usually more than one or two frequencies are generated. For these reasons we have to guess at what size capacitor to use and test its effectiveness in killing the radio noise problems. Start with a capacitor between 0.1 µF and 0.01 µF. When you buy capacitors for this purpose, pay attention to the voltage ratings. Your bot's system may be 24 volts, but voltage spikes that are much higher can be produced. Try to find capacitors that have a voltage rating of at least three times the motor voltage or more.

Now that you know you need a capacitor, you need to know what to do with it. There are a couple of different things to try, depending on whom you ask. One group says to use three capacitors per motor, as shown in the right half of **Figure 13.9**. You can see that this is done by placing a capacitor between the brushes and one from each brush to the motor housing. The other group says to use only one capacitor, between the brushes, as shown in the left half of **Figure 13.9**. It is claimed that you can introduce a high current directly to the frame of your bot if you use three capacitors. I have done it both ways in different bots and haven't seen that problem personally. There are some builders who have skipped using capacitors and had no RFI problems. If you go that route, remember the capacitor when your bot starts freaking out.

Figure 13.10 should give you a good idea about the mechanics of mounting capacitors to a motor. I cannot cover every type of motor, but the main thing to remember is that the

Figure 13.9

Schematic for connecting capacitors to motors.

 OR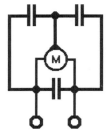

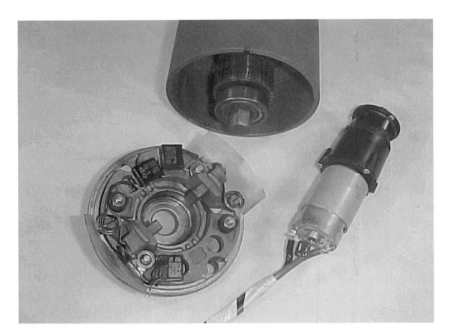

Figure 13.10

Installed motor capacitors.

capacitors should be as close to the brushes as possible. If you cannot open the motor casing, go ahead and install the capacitors on the outside. **Figure 13.10** shows two examples of motor capacitor mountings. The first example shows the motor case opened to install the capacitors. This is a fairly large motor and the caps can be mounted directly on the brush housing. The second example shows where a single cap was mounted on the outside of a cordless drill motor.

There are a few other things you can do to battle RFI. You can twist all the power wire pairs. Each motor has a pair of wires leading out. Twisting the wires creates a capacitorlike situation and can help with RFI. Ferrite cores can be used, but it is very difficult to determine exactly where to put them and which size to use. You can route wires through braided steel to help cut down on the amount of RFI that radiates out of them. One of the best—and sometimes hardest—things to do is keep control circuitry away from power-delivering wires. However, you can place the circuitry inside its own metal housing. Aluminum housings are easy to use and lighter, but offer less protection than steel housings.

FCC Regulations

The last thing I have to say about radios concerns the law. Two very popular bands of hobby remote controls use 72-MHz and 75-MHz bands. The 72-MHz band is reserved by federal law for remote controlled aircraft. The 75-MHz band is reserved for everything else on land or water. Using a remote control system in a manner inconsistent with the laws is punishable by a $10,000 fine. The problem is that the 75-MHz band RC setups don't usually have as many control channels available as the aircraft band RCs. So, for several years, most people who built battling robots used the aircraft band and ignored the law. I have not heard of a case in which the FCC has pressed charges against a builder or an event promoter for using the aircraft band, but now that the sport is gaining immense popularity, anything is possible. It is becoming more popular to buy a 72-MHz system and have it converted to the land frequency. Conversion is a fairly quick and inexpensive procedure if your entire setup can be done, but some receivers cannot be converted. This leaves you two choices if you want to comply with the law: You can have your transmitter converted and buy a new 75-MHz receiver, or you can sell your entire setup and buy one that works on the lawful frequency. It's up to you to decide, but I recommend compliance with the laws.

Speed Controllers

There are three basic ways to control your robot's speed and direction. They are the "bang bang" method, the variable-current method, and the pulse width modulation (PWM) method.

The "bang bang" method isn't really a method of speed control. The motor is either completely on or completely off. This method is the cheapest and easiest way to implement control over a motor. Some people use it to control their robots but it is more suitable to use it to control a weapon motor that doesn't have to turn on and off very frequently. In some cases, as with a spinner motor, you don't need to change the motor's

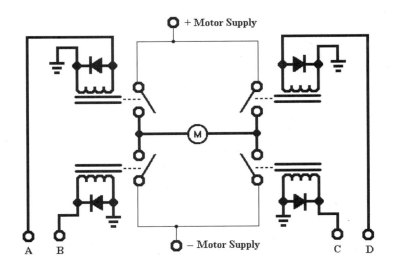

+ Motor Supply

− Motor Supply

A B

C D

Figure 13.11

A simple H-bridge circuit.

spin direction. Turning that motor on and off can be done with a simple contactor. However, some weapon motors, as with a lifter or clamper, need to be turned on and off and have their direction of spin changed fairly often. This requires, at the very least, an H-bridge circuit like the one shown in **Figure 13.11**.

This circuit is made for a large motor. It uses four large contactors to change the voltage to each lead of the motor. Only one pair of contactors is on at a time and these are the ones that diagonally oppose each other. In the schematic they are arranged to look like an "H." When the top contactor of the left side and the bottom contactor of the right side are energized, the motor spins in a clockwise direction. When the bottom contactor of the left side and the top contactor of the right side are energized, the motor spins in a counter-clockwise direction. If you turn on all four, or any two on the same side at the same time, you will melt your batteries and possibly cause a fire. This is a fine method for controlling your weapon motor but it does cause a lot of current to be drawn when you first turn it on. You can use a real speed controller to start the motor out slowly and then gradually speed it up. This will help keep the current draw down to a minimum but you have

to spend time manually controlling it. There is an automatic method for doing this and I'll get to it a little later.

The second method I mentioned was the variable-current method. This method requires some way to vary the current the motor gets. Lower current means the motor will be weaker, therefore it will spin more slowly. The only problem with this method is that the variable resistor that can handle the amount of current from even a small motor that's suitable for a combat bot must be very large. You would also have to have a way to turn the variable resistor remotely.

The best and most popular way to control the speed of a motor is called pulse width modulation (PWM). PWM is a technique of changing the voltage the motor sees by varying the width of a pulse of voltage to the motor. Basically, the pulse is either on or off. While the pulse is on, the motor is trying to run at full power. While the pulse is off, the motor is slowing down and eventually stopping. If you turn the pulse on and off fast enough, the motor's electrical characteristics fool it into thinking it is receiving a lower voltage. That makes the motor run slower. The drawback to PWM is that the controllers that do this technique are fairly expensive. Also, they are more fragile than other types of controllers. When using a PWM controller, be sure to protect it from vibration and impact.

Several brands of PWM speed controllers on the market are currently being used successfully in combat robotics. Each has advantages and disadvantages. 4QD (www.4qd.co.uk) is a British company that makes several controllers that use between 12 and 48 volts. They are capable of controlling a range of currents between 0 and 300+ amps. They also have a lot of features that might come in handy when used on a bot. Vantec (www.vantec.com) is a California-based company that makes several controllers that use voltages ranging from about 5 to 60 volts and are capable of controlling currents ranging up to about 220 amps. They have a built-in mixing function that helps to control your bot. Innovation First Inc. (www.ifirobotics.com), a Texas-based company, also builds a couple of speed controllers that are very popular. The Victor

and the Thor can run at between 12 and 24 volts. The Victor is rated at 60 amps, while its big brother Thor is rated for 120 amps. They are planning 42-volt and 48-volt versions of the Victor and Thor, respectively. Robotpower.com is offering the Open Source Motor Controller (OSMC). Several motor controller experts within the combat robotics community designed the OSMC. It has features like a built-in mixer, flipped bot motor reversal input and a serial interface. It can handle up to about 160 amps and 50 volts. Since it is open source material, if you felt like it you could download the schematics and build your own.

Whether you are using a speed controller to move your bot or move your weapon, there are some things you should do to help keep this expensive piece of equipment in working order. The main thing is to keep the electronics insulated from vibration. Components, while soldered in, can vibrate loose and cause a failure. Some controllers come with half or even less of an enclosure. If this is the case, you should build one or adapt an existing one so that flying metal, grease, or other debris doesn't get into the boards. Follow the manufacturer's ratings. If it says 100 amps is okay, then don't try to squeeze in 120 amps. If it is rated for 24 volts, you had better not try to run it on 36 volts.

Batteries

There are certain battery types that are not allowed in robot combat. The main one, and probably the most surprising one to newcomers, is the regular lead acid battery found in cars, motorcycles and lawnmowers. Anything with liquid electrolyte (that stuff inside the battery that burns your skin) is not allowed in combat. If you hit that type of battery with a hammer or a spinner going 200 miles per hour, you will certainly spray everything in the general vicinity with acid. Currently there are three types of battery in wide use on fighting bots.

The first type is the sealed lead acid (SLA) or gel cell. This sounds like it would violate the rule I just mentioned, but in

reality it doesn't. SLAs do not have liquid electrolyte. They have a gel substance that does not leak when the battery is cracked. The most popular brand of SLA is the Hawker or Odyssey brand. Hawker actually makes the Odyssey brand and packages it for sale to personal water craft and motorcycle drivers. The one thing that makes the Hawker the best brand is that it can supply a lot of current in a short period of time without damaging the battery. One feature of the Hawker is the ability to take in lots of current when charging. There is no limit on the charge current as long as you can keep the voltage at the prescribed level. However, there are two main drawbacks to this type of battery. The worse is the fact that the batteries are heavy. A battery that can supply 90 amps for 5 minutes weighs about 12 pounds. The other drawback is that they don't have a very flexible voltage level. For the most part, Hawker batteries are limited to 12 volts. You must add another 12-pound battery to increase the voltage to 24 volts. It will get very heavy if you want to run 36 or 48 volts at more than about 90 amps.

The second and third types of batteries used are the nickel–cadmium (NiCd) and Nickel–Metal Hydride (NiMH) batteries. These batteries resemble the normal alkaline batteries that you might put in a flashlight but are rechargeable. They have high current output capability, and are also lighter and more flexibly mounted than SLAs because they come in 1.2-volt cells. Because they come in cells with smaller voltages, it is much easier to customize the voltage output and even the weight of a NiCd or NiMH battery pack. One of the drawbacks of these types of batteries is price. Cells can cost between two and eight dollars or more. The chargers you need cost a bit more than an SLA charger too, since they are more complex.

Electronic Interfaces

So far, I've been telling you about some magical radio interface that converts a standard radio signal into control commands for relays, contactors, and whatever else you need to turn on

and off to control your bot. There is even one that I said would turn higher voltage into a voltage suitable for your receiver. Of course you know they aren't magic. The following devices are available at the Team Delta Web site (www.teamdelta.com).

Dan has designed all these devices specifically for the combat bot builder so they are ideal for controlling your new machine. In fact, Dan is a bot builder himself. That is how his devices came to be. He got sick of dealing with glitchy analog circuits with reed relays so he developed the RCE200 line of solid-state switches. They are designed in every way to fail safely. "Even if you overcurrent the switching FET, it explodes and opens the circuit, as opposed to melting shut." The NBC TV network special effects guys that built ChinKilla for Jay Leno use them to fire pyro effects since they're so stable and reliable. They're even stocked at a hobby shop in Burbank for the movie effects crowd; in fact that's where most of them go. His Web site is full of other goodies like wires, switches, connectors, motors, wheels, axles, and battery charger supplies.

Dan has one device that I won't completely cover here. It is still in development and not on the market yet. The RCE510 Digital RX Simulator is his newest project. Only the IFI robot controller has the ability to let builders turn their transmitters and receivers on and test their bots without fear of accidentally controlling another bot across the pit area. Dan's device will allow builders to plug their speed controllers or other devices in and simulate signals from the receiver without broadcasting actual transmitter signals. It also has a mode that cycles between full-forward and full-reverse stick positions. You can turn a knob to control the cycling speed.

The Battery Eliminator Circuit (Part # RCE86)

The battery eliminator circuit (BEC) is used for supplying voltage to your radio receiver. Lots of people use the standard four-cell NiCad battery pack for this but if you start controlling large relays you can quickly drain those batteries and be left sitting idle in the middle of the arena while your opponent

chews on your bot's leg. Instead of living that nightmare, you can use one of the two kinds of BECs that Dan has in stock. The first one is the RCE85. It is a little smaller in that it supplies 300 ma. I'll actually be covering the RCE86 which supplies up to 1.2 amps. It is more powerful because of a few other devices, specifically the bigger dual ended switch. It takes a lot of current to run the two large relays.

Figure 13.12 shows the datasheet for the RCE86 BEC available for download on Dan's Web site. There are three different versions of the BEC. Each has the same output voltage of 5 volts and output current of 1.2 amps. However, each version has a different input voltage range. The RCE86-12 should be used on 12-volt systems but can be used on systems ranging from 9 to 18 volts. The RCE86-24 can be run on 18–36 volts and the RCE86-48 can be run on 35–72 volts.

Figure 13.13 shows the RCE86-24 in use along with a simple schematic. Remember that the RCE86-24 runs off 24 volts. The other versions of the RCE86 and the two versions of the RCE85 all connect to the battery and the receiver in the exact same way. There are four simple connections that you have to make to get this board running. First, you must connect it to the supply battery and then you have to hook it up to the radio receiver. Power connections are easily distinguishable since the solder pads are shaped differently to indicate their polarity. You can cut the wires off an old servo and solder them to the output side of the BEC to make it easier to connect to your radio. All of this is very easy. Dan even gives you center-hole measurements so you can drill mounting holes in your enclosure.

Remember that one of the advantages of a BEC in your system is ground isolation. When motors are connected to batteries, the electrical noise they generate can travel through the power wires back to the battery. This noise can cause problems with radio reception and operation. The BEC will act as a wall that blocks this noise from your radio receiver. You just have to make sure you don't connect the receiver ground directly to the battery ground.

RCE86 BIGGER BATTERY ELIMINATOR · INSTRUCTIONS

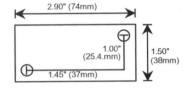

CONVERTER SPECIFICATIONS

Input voltage	RCE86-12 (12 volt) 9 to 18 volts
	RCE86-24 (24 volt) 18 to 36 volts
	RCE86-48 (48 volt) 35 to 72 volts
Output voltage	5 vdc (80mV ripple P-P)
Output current	1200 ma

Figure 13.12

RCE86 datasheet. (Courtesy of www. TeamDelta.com)

MECHANICAL MOUNTING

The RCE86 PC board was designed to be secured by two #6 machine screws to a flat surface. The orientation is not important. If you use small stand-off spacers be sure they do not short any of the traces on the board. If you intend to use this product in an environment subject to harsh shocks you may wish to locate some rubber vibration isolators to use in place of stiff metal stand-offs.

2.90" (74mm)
1.00" (25.4.mm)
1.50" (38mm)
1.45" (37mm)

WIRING HOOK-UP

You will only need to solder four wires to get this industrial DC-DC converter installed in your system. Please refer to the description and the diagram below:

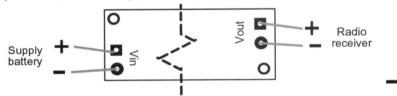

Supply battery

Vin

Vout

Radio receiver

"Vin" connections:
• Square pad to the positive side of your main supply battery.
• Round pad to battery ground

"Vout" connections:
• Square pad to the +5 volt line of your R/C receiver
• Round pad to receiver ground

If you are unsure as to the correct R/C receiver wires to use, consult the operator's manual that came with your R/C rig.

Hint: cut the interface wire off an old airplane servo and solder the +5 and ground lines to Vout pins to make it easy to plug into your receiver. Or, add an optional R/C servo lead to your order.

 Important!

If you connect the battery supply ground to the radio receiver ground you will lose out on significant noise reduction advantages in your system. The RCE86 will still work in this application but it is not optimal.

Figure 13.13

Using an
RCE86-24.

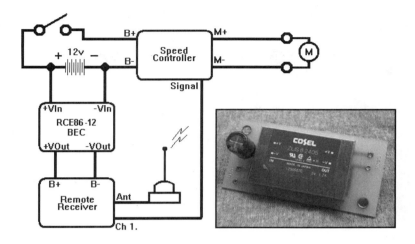

Digital Switches (Part # RCE200X)

The next two devices are the simplest devices in the lineup, not to mention the most flexible. All but one of the rest of the devices we'll talk about are built on this one concept: to give the builder a way to decode signals from the receiver to turn another device on and off. There are four different types of switch. The "X" designation on the part number represents which type you want. The "A" version can switch up to 20 volts and 4.5 amps. The "B" version can switch up to 60 volts and 2 amps. The "C" version can switch up to 20 volts and 9 amps. The "D" version can switch up to 60 volts and 4 amps. Mainly you will use one of these devices to turn a contactor or relay on or off. I've used them to operate flashing lights and other cosmetic effects on bots. In small bots, probably a few pounds or less depending on the size of the motor, you could use it alone to turn on a spinner or lifter motor. Just make sure to strictly follow the maximum voltage and amperage ratings.

Mounting for this device is very simple. It is light enough to be secured to any nonconductive surface inside your bot. You can wrap it in electrical tape or shrink wrap and let it dangle if you want. I wouldn't recommend that. You should at least tape or strap it to a frame member. Just as with the BEC, you can cut the wires and connector off an old servo and solder the leads to the board. Again, Dan has shape-designated the

RCE200x SERIES DIGITAL SWITCHES

MOUNTING

The RCE200 digital switch (D-switch) is small and light enough to be taped to a non-conductive surface or lashed with a tie wrap. A 3/4" piece of clear heat shrink tubing makes an excellent insulative cover.

DIMENSIONS

Width: 0.6" (15mm)

Length: 1.6" (41mm) (A,B models)
2.0" (51mm) (C,D models)

Figure 13.14

RCE200X Datasheet (Courtesy of www.TeamDelta .com)

HOOK-UP AND CONFIGURATION

You will need to solder a three-wire connector suitable for your R/C receiver to the left side of the RCE200 board. The standard colors of the conductors vary between manufacturers so double check their functions before proceeding. The top of the board has all of the components on it. The power pad is the only square pad and toward the top.

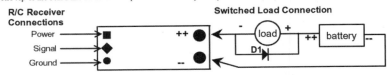

The right side of the board connects to the load you wish to switch on and off. You must connect the D-swich as shown or the switch will not appear to operate properly. Do not exceed the voltage or current specifications of the board or you will destroy it!

Diode D1 is optional but recommended if the load you are switching has high inductance, like a relay coil or motor. The diode will help absorb high voltage transients that often occur when these devices are switched off. A 1N4001 diode is acceptable.

OPERATION

The onboard LED will aid in switch setup and display the status of your radio link:

○ Off   The board is unpowered

● On solid   Transmitter fault: no vaild signal detected (switch is **OFF** for safety)

✦ Slow blink Valid signal; switch is **OFF**

✦ Fast blink Valid signal; switch is **ON**

Two DIP switches set the transmitter stick threshold for turning the switch on. You may alter the switch configutation "on the fly" to test out the various thresholds. You may need to adjust the trim on your transmitter stick for best operatior

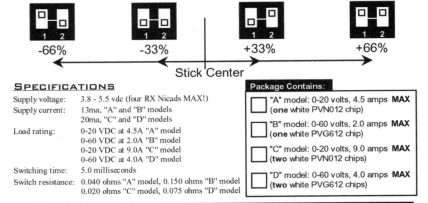

-66%   -33%   +33%   +66%

Stick Center

SPECIFICATIONS

Supply voltage: 3.8 - 5.5 vdc (four RX Nicads MAX!)

Supply current: 13ma, "A" and "B" models
20ma, "C" and "D" models

Load rating: 0-20 VDC at 4.5A "A" model
0-60 VDC at 2.0A "B" model
0-20 VDC at 9.0A "C" model
0-60 VDC at 4.0A "D" model

Switching time: 5.0 milliseconds

Switch resistance: 0.040 ohms "A" model, 0.150 ohms "B" model
0.020 ohms "C" model, 0.075 ohms "D" model

Package Contains:

☐ "A" model: 0-20 volts, 4.5 amps **MAX** (**one** white PVN012 chip)

☐ "B" model: 0-60 volts, 2.0 amps **MAX** (**one** white PVG612 chip)

☐ "C" model: 0-20 volts, 9.0 amps **MAX** (**two** white PVN012 chips)

☐ "D" model: 0-60 volts, 4.0 amps **MAX** (**two** white PVG612 chips)

solder pads to make it nearly impossible to get the wiring mixed up. On the other side of the board are the positive and negative solder pads for whatever load you plan to turn on and off. You have to make sure you connect the negative end of the battery to the negative solder pad. In the data sheet, notice the diode across the load leads. This is the same diode we discussed earlier. It removes spikes generated by motors or relays.

This device has several features. The most important is actually implemented in all of the Team Delta devices. That feature is the failsafe mechanism. Through programming, if the radio signal from the receiver or transmitter is lost, the device will turn off whatever load you are switching. Another feature is the onboard LED. This helps you set up the switch and displays the status of your radio link. When the LED is off, the board has no power. When it is on all the time, there is no valid signal detected from the transmitter, and the switch is off for safety reasons. When the LED is blinking slowly, there is a valid signal from the transmitter and the switch is off normally. When the LED is blinking quickly, there is a valid signal and the switch is on normally. The last feature is that the joystick threshold is adjustable. That means you can determine how far and what direction you have to push the stick to get the switch to turn on. Two small DIP switches accomplish this task and can be switched while the board is powered so you can determine your most comfortable setting while actually running the load.

In **Figure 13.15** I show a schematic that uses four RCE200C devices to run a simple H-bridge like the one I mentioned in **Figure 13.11** earlier in the chapter. You can use some fairly big relays in this circuit since the RCE200C can handle up to 9 amps. Many people use simple automotive relays with no problems. Just make sure your relay contacts can handle the amps used by the motor you are controlling. If you had relays that drew more than 9 amps, you could always use the smaller relay to turn on the larger relay. Then the large relay could control the motor. The only problem with that is that you are

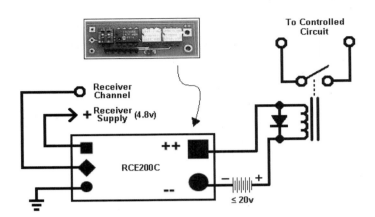

Figure 13.15

Using an
RCE200C.

now venturing into a lot of wiring that could be difficult to troubleshoot while at an event.

Looking at the figure again, you should notice something else really cool about these boards. You can wire them up to the same receiver channel and control more than one at a time. If you look at the datasheet you can see that these boards can draw up to 20 ma while running. So make sure your radio receiver battery pack can source the total amount of current, 80 ma in this case. 80 ma isn't too much for most packs, but once you start stacking other devices on the receiver, the power demand can grow quickly. That's when the BEC comes into play.

If you recall from the simple H-bridge circuit, you should never turn on the two inputs on the same side at the same time. Look at **Figure 13.15** again. Imagine that each contactor is controlled by an RCE200C. The DIP switches for the diagonally located RCEs are set to activate at the same stick distance and direction. The other two diagonally located RCEs are set to activate at the opposite stick distance and direction. When they are wired and set up this way, it is impossible to turn on the incorrect inputs at the same time. If I were you, I would try controlling a small motor and small battery so that I didn't burn up too many relays and batteries before I knew I had it right.

Relay Switch (Part # RCE210)

Next up we have the relay switch. This part is almost exactly the same as the digital switch. The difference is that a standard automotive relay is mounted to the circuit board. You don't have to concentrate on getting it to work like you do with the RCE200X and there are fewer less solder joints that you have to get right. In fact, the relay has two 1/4-inch spade connectors. You use these to connect the large load that you'll be switching.

Mounting is done in basically the same manner as in the RCE200X. Add an old servo lead to the shape-designated solder pads, connect your motor via spade connectors to the relay, and wrap it up so that none of the connections on the bottom will short out on your frame. Then strap it down to a frame member. However, if you are neater than that, there are two mounting holes on the board designed for #6 machine screws. Use the screws with a couple of standoffs and you have a very tidy solution. Just make sure that you support the board beneath the relay when plugging in the spade connectors. If you don't, you could possibly crack the circuit board.

If you look at **Figure 13.16**, the datasheet, you will see that this device has the same features as the RCE200X. It has the single LED that lets you know the status of the board and radio signal. It has the DIP switches that can be set so that different stick positions and directions can be used to turn the motor on and off. The biggest difference, again, is the relay. When using the RCE200X, you must supply the relay's power source. With the RCE210, the relay gets its power directly from the radio receiver battery pack through the receiver connection. When the relay is turned on, the board draws 220 ma from the battery pack. This is like running twelve of the RCE200X boards. The small receiver battery pack can supply this current for a while, but you do run the risk of draining it quickly. This is really apparent when you run more than two RCE210s. So far I've only *suggested* the use of a BEC. Now it is time to *recommend* its use.

RCE210 RELAY SWITCH

MOUNTING

The RCE210 has diagonal mounting holes for #6 machine screws. Alternately you may wrap the entire unit in some form of non-conductive padding and securely zip-tie it in place.

HOOK-UP AND CONFIGURATION

You will need to solder a three-wire connector suitable for your R/C receiver to the top left side of the RCE210 board. The standard colors of the conductors vary between manufacturers so double check their functions before proceeding. The top of the board has all of the components on it. The power pad is the only square pad and toward the right.

The black cube is the relay that actually switches your load on and off. It has 0.25" spade connectors for easy attachement of wires - no soldering required! Take care in supporting the bottom of the RCE210 when pushing connectors onto the relay to prevent flexure and possibly breakage of the PC board.

OPERATION

The onboard LED will aid in switch setup and display the status of your radio link. The blink codes are standard for all Team Delta interfaces and are as follows:

○ Off   The board is unpowered

● On solid   Transmitter fault: no vaild signal detected (relay is **OFF** for safety)

PHYSICAL

Size: 1.75" x 2.50" x 1.25"
Weight: 4 oz.

R/C Receiver Connections

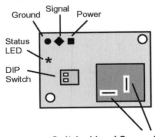

Switched Load Connectors

✷ Slow blink   Valid signal; relay is **OFF**

✷ Fast blink   Valid signal; relay is **ON**

Two DIP switches set the transmitter stick threshold for turning the relay on. You may alter the switch configutation "on the fly" to test out the various thresholds. You may need to adjust the trim on your transmitter stick for best operation.

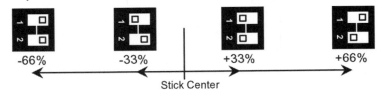

Stick Center

SPECIFICATIONS

Supply voltage:	4.1 - 5.5 vdc (four Rx Nicads MAX!)
Supply current:	20ma stand-by 220ma when relay energized
Load rating:	Relay is rated for non-welding contact up to 24 amps at 30 volts DC
Switching time:	8.0 milliseconds typical

✋Important!

This product is a current hog. You might not think that 220 milliamps is a lot, but to a AA size Nicad receiver battery pack, it is.

BE DILIGENT in keeping those receiver batteries topped off. The relay in this product drops out at 4.1 volts, so it really needs a good, strong power source to operate properly.

Figure 13.16

RCE210 Datasheet.

Figure 13.17 shows the basic schematic and a picture of an RCE210 in use, controlling a small motor. Notice that you still have to supply the power for the motor itself. You must also take into account the contact ratings of the relay. This particular relay is rated for up to 24 amps at 30 volts. If you use a motor that draws more than 24 amps, you run the risk of burning the contacts in the relay. Depending on how many more amps you are pulling, the relay could slowly die or it could burn up almost instantly. Just as with any other part of your bot, you are the one who decides how far to push current ratings. You wouldn't run a heavyweight bot's spinner weapon directly with this relay, but it might be possible in a 12- or maybe even a 30-pound bot. You should check what your competition is doing with their bots though. You might be overpowered in the ring.

Figure 13.17

Using an
RCE210.

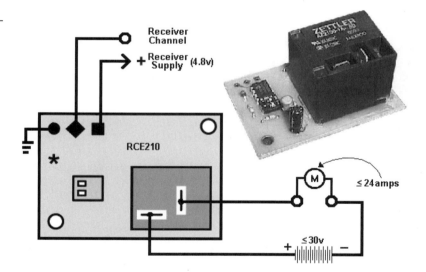

Bigger Dual Ended Switch (Part # RCE225)

The bigger dual ended switch is another device that was designed as an upgrade to the simple RCE200X. This time two standard relays have been added to the board along with a couple of functionality surprises.

RCE225 BIGGER DUAL ENDED SWITCH - INSTRUCTIONS

Figure 13.18

RCE225 Datasheet part 1. (Courtesy of www.TeamDelta .com)

1.90" (48mm)
2.75" (70mm)
2.40" (61mm)
3.25" (83mm)

MOUNTING

The RCE225 PC board was designed to be secured by four #6 machine screws to a flat surface. The orientation is not important. If you use small stand-off spacers be sure they do not short any of the traces on the board. If you intend to use this product in an environment subject to harsh shocks, take care to affix the shorting jumper (below) so it does not fly off.

HOOK-UP AND CONFIGURATION

First you will need to solder a three-wire connector suitable for your R/C receiver to the top left side of the RCE225 board. The standard colors of the conductors vary between manufacturers so double check their functions before proceeding. The top of the board has all of the components on it. The role of each solder pad is keyed: square for power, diamond for signal and round for ground. If you reverse power and ground you will destroy this board, so double check your work.

R/C Receiver Connections

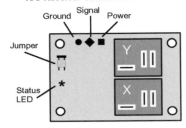

Ground Signal Power
Jumper
Status LED

Switched Load Connections

Each relay has its own contact connections for controlling your device. There are both normally-open (N.O.) and normally-closed (N.C.) pads available for maximum flexability. Each relay operates independently, though an electronic safety interlock prevents both relays from becoming energized at the same time.

Two pairs of connections are available at the bottom of the board that allow external limit switches to override the R/C transmitter stick command. To force Relay Y into the de-energized state simply short the two pins labled "LimY." Relay X has the same provision for it as well.

OPERATION

The red LED will aid you in setup. It's operation is standard on all Team Delta products:

○ **Off** — The board is unpowered

● **On solid** — Transmitter fault: no vaild signal detected (all relays **OFF** for safety)

✹ **Slow blink** — Valid signal; all relays are **OFF**

✹ **Fast blink** — Valid signal; one relay is **ON**

The sensitivity jumper configures the Relay X and Relay Y turn on/off thresholds as shown in the table. 0% is stick center. You may alter the jumper configutation "on the fly" to test out the two thresholds. You may also need to adjust the trim on your transmitter stick for symmetrical operation.

	Xon	Yon
Jumper installed	-66%	+66%
Jumper removed	-36%	+36%

RCE225 Specifications

Supply voltage:	3.8 - 5.5 vdc	Relay load rating:	30VDC 24 amp motor load
Supply current:	20ma static, 220ma energized	Switching time:	8 milliseconds (typ.)

Check out the datasheet of the RCE225 shown in **Figure 13.18**. You'll see that there are four, #6 mounting screw hole in the board. This device is a little too heavy to simply strap to a frame member. The weight of the board could cause it to crack when your robot slams into its opponent. It is suggested that you use short standoffs and the screws so that it is securely mounted. There is also a jumper instead of a set of DIP switches. If your design has the jumper in place, you should secure it with a drop or two of hot glue.

The jumper is used to set the "on" position of the relays. With the joystick centered and the jumper installed, the "X" relay will turn on when the joystick is positioned past the negative 66 percent mark. Without the jumper, the "X" relay will turn on when the joystick is positioned past the negative 36 percent mark. The "Y" relay turns on, according to the jumper, when the joystick is positioned in the positive side of movement. This means that there is no way to turn both relays on at the same time, but they do operate on the same receiver channel. This will make it easier to construct a low- to medium-current H-bridge circuit as shown on the right side of **Figure 13.19**. You can use two of these boards to have the "bang bang" control of a small bot. You can also use them to control two motors at different times on the same channel.

The RCE225 has a feature that isn't included in the single-relay or no-relay versions of the device. This device has inputs for "X" and "Y" relay limit switches. A limit switch is used to signal that a linearly moving device has reached its mechanical limit. In many robots, builders have used linear actuators to push some type of mechanism. This can take the form of a clamping device, a self-righting device, or whatever else you might think of. **Figure 13.20** shows the RCE225 put to use as a controller for a linear actuator. It is wired up as a simple H-bridge controller. There are two microswitches, hooked up respectively to the "LimX" and "LimY" connectors.

To force Relay Y into the de-energized state, simply short the two pins labeled "LimY." Relay X has the same provision. When the joystick has Relay X on, the actuator is pushing out.

BIGGER DUAL ENDED SWITCH

Figure 13.19

Using an
RCE225 as an
H-Bridge.
(Courtesy of
www.TeamDelta
.com)

RCE225 - DESIGN NOTES

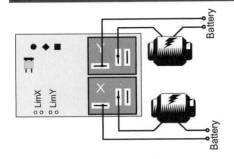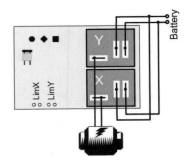

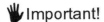
To control two
different motors

To control one motor
in forward and reverse

✋ Important!

This product is a current hog. You might not think
that 220 milliamps is a lot, but to a AA size Nicad
receiver battery pack, it is.

BE DILIGENT in keeping those receiver batteries
topped off. The relays in this product drop out at
4.1 volts, so it really needs a good, strong power
source to operate properly.

Figure 13.20

Using an
RCE225 to
control a linear
actuator.

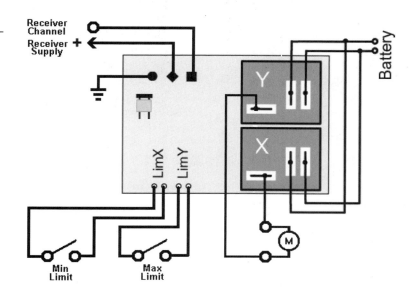

In **Figure 13.21** you see that when the actuator reaches its extreme limit, limit switch X will close. This will short the "LimX" pins, turning off the motor. When you move the joystick to the opposite side, Relay Y will turn on. This will make the actuator pull back. Once the actuator hits its mechanical limit, limit switch Y closes. Now the motor cuts off and waits for you to move the joystick again. Partial movement of the actuator is accomplished by putting the joystick back in the

Figure 13.21

Using limit
switches.

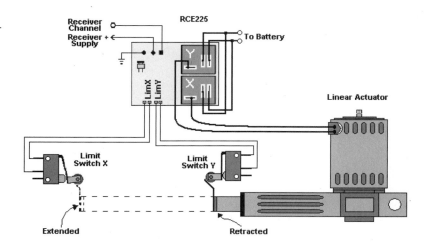

center position before the actuator reaches the limit. **Figure 13.21** shows the mechanical stop attached to the actuator arm itself. I did it this way for simplicity. Normally the mechanical stop will be on whatever part the actuator is moving. Or, the part that is moving could be the part that closes the switch. It is up to you to design for using limit switches.

RC Contactor Set (Part # RCE218)

This new device is called the RC contactor set for a reason. It pairs the RCE200B digital switch with a large, continuous-duty, White Rogers solenoid (or contactor). You use this device to control large motors. Before Dan started stocking the contactor set, I used the RCE200X switch to turn on the exact same contactor in several bots. I used the combination to switch on motors that drew up to 750 amps at stall.

750 amps, if you look at the datasheet in **Figure 13.22**, is a little over the current rating for the contacts of this device. However, I doubt that much current ever went through it. The motor I was using powered a spinning bar. It had a very high gear reduction so it spent very little time near stall when starting up. The actual ratings for the contactor's contacts are 200 amps continuous and 600 amps for a short amount of time.

Notice in **Figure 13.23** that you still have to supply a power source for the contactor coil. The contactor comes pre-mated with the RCE200B digital switch. The difference in this case is that Dan has already added the servo wire to the circuit board and wired up the contactor. All you have to do is hook up the coil supply and the motor leads and plug in to your receiver. In **Figure 13.23**, I happen to run the weapon motor on 24 volts and use the same supply to power the contactor coil. You do not have to run your contactor off the same supply as your motor but if you do, Dan offers three different versions of this set. Whether you run your weapon on 24, 36, or 48 volts, there is a contactor set that can operate on that voltage. You just have to specify it when you place your order. If you happen to have several RCE200X digital switches in your parts box, Dan will even let you order the contactor by itself.

Figure 13.22

White Rogers Contactor Datasheet.

WHITE-RODGERS DC PRODUCTS

Type 586

sealed solenoid

FEATURES

- Water resistant
- Double-make or double-break contacts
- Capable of handling high and low current requirements
- Black impact phenolic casing

ENGINEERING DATA

Contacts

- Pole form–SPNO and SPDT
- Material–silver alloy
- Termination–⁵⁄₁₆"-24 UNF-2A thread

RATINGS				
	NO		NC	
Volts DC	Cont.	Inrush	Cont.	Inrush
6	200A	600A	100A	300A
12	200A	600A	100A	300A
24 & 36	200A	600A	100A	200A

Coils

- Voltage–6 VDC through 48 VDC
- Termination–#10-32 UNF-2A thread
- Power (approximate)
  Continuous 12 watts SPDT, 8 watts SPNO
- Connections
  Coil isolated (two terminals)
- Operate (77°F/25°C)
  67% of nominal coil voltage (intermittent)
  75% of nominal coil voltage (continuous)
  110% max. safe of nominal coil voltage

COIL DATA				
	Resistance in Ohms			
	Intermittent§		Continuous	
Volts DC	SPDT	SPNO	SPDT	SPNO
6	2.2	3.3	3.3	5.25
12	8.4	13.2	13.2	21.0
15	13.1	21.0	21.0	32.8
24	18.4	30.0	30.0	47.0
28	33.6	53.0	53.0	84.0
36	75.6	120.0	120.0	189.0
48	134.0	213.0	213.0	336.0

§ Intermittent—special request only

GENERAL DATA

Dielectric Strength
- 500 Volts

Temperature Range
- -40°F/-40°C to 149°F/65°C

Mechanical Life (no load)
- 100,000 cycles

Electrical Life (rated load—making & breaking 200 amp on NO Contacts)
- 50,000 cycles

Mounting Position
- Recommended mounting is vertical plane with coil terminals up

Weight (approximate)
- SPNO–24.0 oz.
- SPDT–26.0 oz.

Duty Cycle
- Continuous
- Intermittent—10 seconds "on" maximum and minimum 60 seconds "off"
  One minute "on" maximum and minimum 6 minutes "off"

Hardware Torque Specification
- Contact Terminal: 60 inch-lbs. max.
- Coil Terminal: 12-18 inch-lbs. max.

WHITE-RODGERS

White-Rodgers Division, Emerson Electric Co.
9797 Reavis Road, St. Louis, MO 63123-5398

www.white-rodgers.com

R-4005
0004

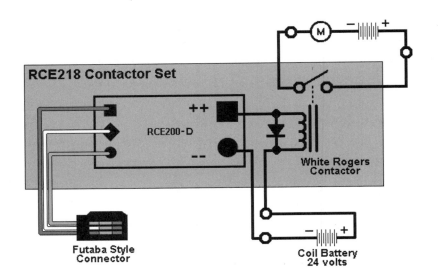

Figure 13.23

Using an RCE218 to control a large weapon motor.

Digital Servo Slower (Part # RCE550A)

The name of this last device is a little confusing, if you ask me. However, it does do what the name says. You just won't normally use it to slow a servo. You'll use it to start a weapon motor out slowly so that it doesn't draw as much current as it would if you just used a contactor to turn it on. All you have to do is connect it to your radio and speed controller, select the time delay, and it works.

The datasheet for the servo slower, shown in **Figure 13.24**, is fairly short. That's because using the device is fairly simple. The device comes prewired with a Futaba-style radio connector so that you can plug it directly into your receiver. If you don't have a Futaba radio, and the wires do not match up, it is usually a simple matter of changing the wire position in the connector to get it right. It also comes with a Futaba-style connector to plug into your speed controller. Mounting is accomplished in the same way as for an RCE200X. Insulate it and strap it to a frame member or place it inside a component box along with the rest of your electronics. It has a status LED too. When the LED is off, the board is unpowered. When on solid, the board is powered but there is no signal being received

Figure 13.24

RCE550A
Datasheet
(Courtesy of
www.TeamDelta
.com)

RCE550A - DIGITAL SERVO SLOWER

MOUNTING

The RCE550A digital servo slower is small and light enough to be taped to a non-conductive surface or lashed with a tie wrap. A 3/4" piece of clear heat shrink tubing (included) makes an excellent insulative cover.

DIMENSIONS

Width: 0.6" (15mm)
Length: 1.6" (41mm)

HOOK-UP AND OPERATION

The RCE550A comes prewired with Futaba style connectors. All you have to do is plug it in and select the time delay you want!

Operation is simple: when you first power-on the RCE550A, it immediately passes the input signal to the output. From then on, it slows the input signal from the R/C receiver to the output device by whatever full-scale time range you set. In the event of an invalid or missing signal, the RCE550A doesn't generate any signal at all allowing the device being slowed to engage its own failsafe mechanism. When a valid signal returns it slowly tracks to that new setting.

Other than the exception for the *first* instance at power-on, the RCE550A will slow all transitions it receives from the R/C receiver. For instance, if you have the Tx stick at +100%, turn off the Tx, move the stick to -100% and then turn the Tx back on, the RCE550A will smooth the transition from +100% down to -100%.

This interface is very useful when using the IFI Robotics "spin controller" products. Tying the RCE550A to a Tx channel with an on/off toggle switch yields a useful human interface to controlling a high power motor while limiting the amount of current consumed at low RPMs. In addition, the RCE550A obviates the need for a PWM buffer to the IFI Robotics products. This interface also works well with the Astroflight 207D and 212D products.

OPERATION

The onboard LED aids in setup and displays the status of your radio link:

○ Off   The board is unpowered   ✦ Slow blink   Valid signal; slowing active

● On solid   Transmitter fault: no vaild signal
              detected (no output signal emitted)

Two DIP switches set the full-scale time delay between Tx signal extremes, ie -100% to +100%. You may alter the switch configuration "on the fly" to test them out and see what works best in your application.

3.5 sec   4.5 sec   7.5 sec   13 sec

SPECIFICATIONS

Supply voltage: 3.8 - 5.5 vdc (four RX Nicads MAX!)
Supply current: 7ma
Granularity: 200 steps full scale

from the transmitter. When the LED is blinking slowly, there is a valid signal from the transmitter and the device is actively slowing a signal. There is a DIP switch that lets you choose between four different time settings.

When you hook an RCE550A to your receiver, normally you will connect it to the toggle switch output. **Figure 13.25** shows how to use the device along with the toggle switch to turn on a single motor in a single direction. The speed controller here is the IFI spin controller. The spin controller operates a weapon motor in a single direction. Getting an actual spin controller, rather than a standard speed controller, has the advantage of giving you the ability to run more amperage through it.

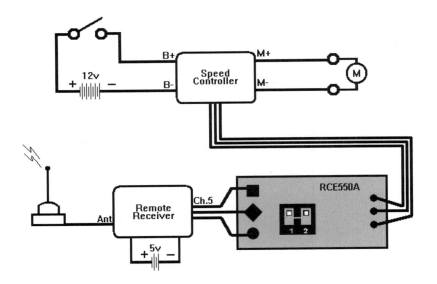

Figure 13.25

Using an RCE550A.

The normal signal from a toggle switch on your transmitter is the same as moving a joystick to the end of its travel, or to 100 percent. Switching it off again is like moving the stick to the opposite end of its travel, or negative 100 percent. So, if you set the time delay to 7.5 seconds, the servo smoother will take 7.5 seconds to ramp up from negative 100 percent to positive 100 percent. Switch the toggle again and the servo smoother

will take 7.5 seconds to ramp down from 100 percent to negative 100 percent.

If you hooked the device up to a joystick channel instead of a toggle channel, it would take 7.5 seconds to ramp from the original joystick position to the new joystick position. In Dan's words…"All it does is put a cap on the maximum rate of change of the channel. The best way to express that cap is as a speed over which the max swing of the channel is smoothed."

This device doesn't have failsafe programming. This one simply forwards any command received, but it does delay them. Whatever device you connect to the RCE550A should have its own failsafe mechanism in place. That won't be a problem with the IFI, Vantec; or OSMC speed controllers.

Summary

This chapter has been all about the radios and electronics involved in combat robot weapons. I covered a little bit about schematics since it is so important that you be able to read them. However, I didn't cover a lot because the ones you need to be able to understand are not that complicated. Wire size is something you need to get right so that you don't cause a fire inside your bot by frying your wires. Following five simple steps outlined earlier should do when starting and stopping your bot. When it comes to radios, there are a few types to choose from. You're better off picking one that costs a little more money and gives you the most flexibility. I also mentioned several radio problems you might face, and their solutions. We also talked a bit about the speed controller and batteries used in fighting bots. The last part of the chapter discussed several radio interfaces that you will find useful in your bot building.

The next chapter will try to give you a glimpse of what is to come in the future of robot weapons and maybe even with the sport itself. There are lots of advances that will come about in weapon systems. There are even more changes that are happening right now in the sport.

14

The Future

I don't have a crystal ball, nor do I possess any extrasensory perception. However, I do hear what some builders would like to see happen in the sport and with the robots themselves. Given enough information, everyone can come to conclusions about what might happen next. That doesn't necessarily mean they, or I, will be right. But there is always a chance.

The Future of the Sport

Earlier in the book I told you my history in this sport. I started out by competing in a small event. That event had about twenty-five or so bots as far as I can remember. Now events are getting twice that many. My next event was BattleBots where the attendance was somewhere around 150. BattleBots was only getting started at that time. Later events saw in the neighborhood of around 450–500 competing robots. Soon there were to be four competitive robotics shows on TV: BattleBots, Robot Wars, Robotica, and a little-known underwater robot competition that may have only been on the air one time. As this book is being written, BattleBots was canceled from TV after its fifth season. Robotica went for three seasons, and

Robot Wars is still running. To my mind, none of this matters a bit.

The real "gold" is in the local events that are springing up all over the country. I've mentioned the event that I organize along with one or two others. However, a new organization called the Robot Fighting League (www.botleague.com) has formed from about twenty events. The RFL is dedicated to getting the sport into the mainstream by raising public awareness and helping new events along by keeping match records, listing arena specifications, and developing a standardized set of tournament rules. For the first time in the history of robotic combat there are enough events around the country to form this league. Plans for the first national event, where winners from other events are the qualifying entrants, have been made, and the events should have happened by the time you are reading this. Some of the qualifying events are even paying for the travel expenses for their winners. Plans for other national collaborations are being laid. In the future, hopefully, there won't be a need for teams to travel all the way across the nation just to compete unless they want to. Even then, we hope that large sponsors will pick up the tab. With a truly organized effort among so many event organizers, this is a very good possibility. The main point is that you should do your best to support your local events even if you don't have a robot to compete.

The Future of Combat Robot Weapons

When it comes to the future of combat robot weapons, there won't be much physical change. There will be small innovations that make a spinner get up to speed faster or a flipper get its opponent higher into the air. But, because of the nature of rules that there be kinetic-energy weapons only, there can only be so many different types of weapon. Everything else will be a variation on existing technology, just as it is right now. Every type of combat robot weapon has its roots in technology that has existed for many years.

The main thing that will happen is the addition of more types of sensors that aid in attacking methods. The gyro is probably the most-used sensor today. Ultrasonic sensors are already in light use as range detectors for weapons that should strike when the opponent is at a certain distance. If your hammer weapon has a reach of 3 feet, an ultrasonic sensor can be used to detect when your opponent is 2 feet away and in the line of fire. Once detected, the hammer can be fired automatically. The same device could be used on a projectile weapon, a spear weapon, or even on a flipper, to make sure the opponent is positioned over the flipping arm.

Another use for sensors will be to monitor the internal environment of the bot itself. Like today's racecar drivers, bot drivers can benefit from knowing internal details of their machines. The temperature and average RPM of drive motors, the temperature and current flow of speed controllers, the temperature and charge level of your batteries, and even the amount of radio frequency noise can all be detected during a practice run or an actual fighting match. There isn't much time during a match to deal with any of this information but it can be saved and downloaded afterwards. Then it can be useful between matches. It could tell you that a drive motor needs to be changed out, or that a speed controller is about to fry.

Some of the really techno-driven builders want to see bots that carry computers with lots of processing power in order to do some really advanced tasks like opponent tracking and near-autonomous movement. I'm a techno-driven builder myself but I don't share the sentiment. I like the way the competition is today. Being a winner means you spend the time to design a superior bot and the time to practice driving it. It will no longer be fun if all you have to do is roll the bot in, push a button, and watch the bots attack each other. In fact, there used to be an autonomous combat robot class. To be fair, building an autonomous bot at all is a daunting task, one that I have been involved in for the last twenty-plus years and will continue to be so. However, that class died out partly because it

was difficult to build the competitors and also because there was so little involvement after you turned the thing on.

The last thing I want to say about the future of robot combat is that what I've written here is my opinion. If you don't agree with it, I don't want to hear it. Instead, prove me wrong with a real working combat robot. I'm not above changing my mind. In fact, I'll probably have a different view of the future by the time this book is printed.

Book Summary

In this book I've tried to give you a summary of what got me to where I am in combat robotics along with what I think might happen in its future. We've talked a little about what is legal and illegal within the existing tournament rules. Be sure to read and understand all the rules of any tournament you plan to compete in. If you have questions, talk with the organizers. They are more than willing to help you compete legally. After all, you are the reason they are holding an event. I've talked a little about getting started. Do it as soon as you can and learn as much as you can. You will find a use for the knowledge you gain through studying the materials, physics, electronics, and mechanics involved in building your bot.

Along with showing what type of bots exist and how they stack up against the other types, I presented some information about their actual structure and what you might want to pay attention to while designing your own. I have given details about batteries and motors too. I also included a good bit about the electronic devices that convert radio signals into bot commands.

I've discussed every type of fighting robot that I can think of. I tried to include a little strategy for competing with each type against each other type. What it all boils down to is the rock–paper–scissors theory of competition. The rock defeats the scissors by smashing them. The scissors defeats the paper by cutting it. The paper defeats the rock by covering it. The

same examples exist among all fighting robots. One type has an opponent that it might easily defeat. That opponent has another type of opponent that it might easily defeat. It is a vicious circle that is only broken by sound bot design, construction, and driving. I guess I could have saved a few trees and just put this last paragraph on my Web site but what would be the fun in that?

Index

Note: Boldface numbers indicate illustrations.

About the CD

The CD-ROM contains CAD and rendering software, CAD models, robot fighting videos and printable tables filled with useful information for all robot builders.

Rhino3D is a computer aided design (CAD) program and provides the tools to accurately model your designs ready for rendering, animation, drafting, engineering, analysis, and manufacturing. (Evaluation version)

Flamingo is a raytracing program that provides photometrically accurate images with reflections, refraction, diffusion, translucency, transparency, color bleeding, shadows, depth of field, depth attenuation, ClearFinish™, and indirect lighting. (Evaluation version)

The robot fighting videos are some of the best fight clips available from the NC Robot StreetFight.

The CAD models include a full representation of Dagoth, the thirty pound bot constructed in *Combat Robots Complete*. This particular model shows Dagoth's upgrade plans from a wedge bot with spike to a vertical spinner bot with a secondary wedge.

About the Author

Chris Hannold has been involved in robotics for almost two decades. A resident of Linwood, North Carolina, he answers robotics questions each day on an Internet forum and organizes and produces the NC Robot StreetFight. He is also the author of McGraw-Hill's *Combat Robots Complete: Everything You Need to Build, Compete, and Win.*

CD-ROM WARRANTY

This software is protected by both United States copyright law and international copyright treaty provision. You must treat this software just like a book. By saying "just like a book," McGraw-Hill means, for example, that this software may be used by any number of people and may be freely moved from one computer location to another, so long as there is no possibility of its being used at one location or on one computer while it also is being used at another. Just as a book cannot be read by two different people in two different places at the same time, neither can the software be used by two different people in two different places at the same time (unless, of course, McGraw-Hill's copyright is being violated).

LIMITED WARRANTY

Customers who have problems installing or running a McGraw-Hill CD should consult our online technical support site at http://books.mcgraw-hill.com/techsupport. McGraw-Hill takes great care to provide you with top-quality software, thoroughly checked to prevent virus infections. McGraw-Hill warrants the physical CD-ROM contained herein to be free of defects in materials and workmanship for a period of sixty days from the purchase date. If McGraw-Hill receives written notification within the warranty period of defects in materials or workmanship, and such notification is determined by McGraw-Hill to be correct, McGraw-Hill will replace the defective CD-ROM. Send requests to:

McGraw-Hill
Customer Services
P.O. Box 545
Blacklick, OH 43004-0545

The entire and exclusive liability and remedy for breach of this Limited Warranty shall be limited to replacement of a defective CD-ROM and shall not include or extend to any claim for or right to cover any other damages, including, but not limited to, loss of profit, data, or use of the software, or special, incidental, or consequential damages or other similar claims, even if McGraw-Hill has been specifically advised of the possibility of such damages. In no event will McGraw-Hill's liability for any damages to you or any other person ever exceed the lower of suggested list price or actual price paid for the license to use the software, regardless of any form of the claim.

McGRAW-HILL SPECIFICALLY DISCLAIMS ALL OTHER WARRANTIES, EXPRESS OR IMPLIED, INCLUDING, BUT NOT LIMITED TO, ANY IMPLIED WARRANTY OF MERCHANTABILITY OR FITNESS FOR A PARTICULAR PURPOSE.

Specifically, McGraw-Hill makes no representation or warranty that the software is fit for any particular purpose and any implied warranty of merchantability is limited to the sixty-day duration of the Limited Warranty covering the physical CD-ROM only (and not the software) and is otherwise expressly and specifically disclaimed.

This limited warranty gives you specific legal rights; you may have others which may vary from state to state. Some states do not allow the exclusion of incidental or consequential damages, or the limitation on how long an implied warranty lasts, so some of the above may not apply to you.